Uttering Trees

Linguistic Inquiry Monographs
Samuel Jay Keyser, general editor

A complete list of books published in the Linguistic Inquiry Monographs series appears at the back of this book.

143

Uttering Trees

106 190

Norvin Richards

The MIT Press
Cambridge, Massachusetts
London, England

For information about special quantity discounts, please e-mail special_sales@mitpress.mit.edu

This book was set in Times New Roman and Syntax on 3B2 by Asco Typesetters, Hong Kong.
Printed and bound in the United States of America.

Library of Congress Cataloging-in-Publication Data

Richards, Norvin.
Uttering trees / Norvin Richards.
    p.   cm. — (Linguistic inquiry monograph ; 56)
Includes bibliographical references and index.
ISBN 978-0-262-01376-5 (hardcover : alk. paper) — ISBN 978-0-262-51371-5 (pbk. : alk. paper)
1. Prosodic analysis (Linguistics) 2. Grammar, Comparative and general—Phonology. 3. Grammar, Comparative and general—Syntax. I. Title.

P224.R53   2010
414′.6—dc22   2009024581

10 9 8 7 6 5 4 3 2 1

Contents

Tagalog, 165–182
Interlude: More wrap 186–

Series Foreword

We are pleased to present the fifty-sixth in the series *Linguistic Inquiry Monographs.* These monographs present new and original research beyond the scope of the article. We hope they will benefit our field by bringing to it perspectives that will stimulate further research and insight.

Originally published in limited edition, the *Linguistic Inquiry Monographs* are now more widely available. This change is due to the great interest engendered by the series and by the needs of a growing readership. The editors thank the readers for their support and welcome suggestions about future directions for the series.

Samuel Jay Keyser
for the Editorial Board

Acknowledgments

I've been extraordinarily lucky, as I've been working out the ideas presented here, to have had the chance to learn from a wide variety of people, and I'm very grateful to all of them for their generosity. Here's a (surely incomplete) list.

The theory in chapter 2 was inspired by Moro (2000) and by Alexiadou and Anagnostopoulou (2001). I'd like to thank Andrea Moro, Artemis Alexiadou, and Elena Anagnostopoulou, not only for writing these works, but for much fruitful discussion since.

The theory in chapter 3 was inspired by a visit to Mornington Island in Australia. I'm very grateful to the Lardil people for welcoming me into their community, and to Ken Hale for giving me the chance to go there.

For help with particular languages and language families, many thanks to Alya Asarina, Željko Bošković, Andrew Carnie, Damir Ćavar, Michel Degraff, Martina Gračanin-Yüksek, Lydia Grebenyova, Kleanthes Grohmann, Ken Hale, Sabine Iatridou, Patrick Jones, Winnie Lechner, Jeff Leopando, Lawrence Maligaya, Victor Manfredi, Joshua Monzon, Pierre Mujomba, Degif Petros, Joachim Sabel, Uli Sauerland, Arthur Stepanov, Sandra Stjepanović, Adam Szczegielniak, Michael Wagner, Susi Wurmbrand, and Kazuko Yatsushiro.

Too many people to count have helped me figure out what I was saying and why; I'd especially like to thank David Adger, Karlos Arregi, John Bailyn, Mark Baker, Željko Bošković, Noam Chomsky, Sandra Chung, Jessica Coon, Danny Fox, Martina Gračanin-Yüksek, Ken Hale, Morris Halle, Claire Halpert, Wayne Harbert, Alice Harris, Sabine Iatridou, Takako Iseda, Shin Ishihara, Hilda Koopman, Richard Larson, Shigeru Miyagawa, Jon Nissenbaum, David Pesetsky, Sasha Podobryaev, Henk van Riemsdijk, Andres Salanova, Lisa Selkirk, Tim Stowell, Esther Torrego, Michael Wagner, Mary Ann Walter, and Hedde Zeijlstra.

I also would like to thank audiences at MIT, SUNY Stony Brook, Middle East Technical University, Boğaziçi University, Indiana University, St. Petersburg State University, Queen Mary University of London, ZAS, the University of Novi Sad, the University of Paris 8, Moscow State University, National Tsing Hua University, the University of Massachusetts at Amherst, Syracuse University, Cornell University, Princeton University, McGill University, the University of Maryland, the University of Illinois, UCLA, the University of Calgary, Rutgers University, the University of Campinas, Yeungnam University, Yale University, the University of Pennsylvania, the University of Minho, and WCCFL 20 at the University of Southern California.

Thanks, too, to Ada Brunstein, Elizabeth Judd, Anne Mark, and Sandra Minkkinen for their superb editing and copyediting, and to four anonymous reviewers.

None of these people are responsible for the many mistakes I am sure I have made, all of which were my own idea.

Finally, many thanks to my family, for their love, support, and grammaticality judgments.

1 Introduction

This is a book about conditions imposed on the narrow syntax by its interface with phonology. The idea that some of the properties of syntax follow from its interface with phonology is not new. Chomsky's (1995) Minimalist Program pursues the idea that most if not all of the properties of syntax are consequences of the need to create linguistic objects that are well suited for the interfaces. Kayne's (1994) Antisymmetry is a particular proposal about an algorithm for rendering the hierarchical structure of a tree as a linearly ordered string of words, which has consequences for which types of trees are acceptable.

I will make two new proposals about the conditions on the syntax-phonology interface. The first, called "Distinctness," is a claim about the nature of well-formed linearization statements, of the type first proposed by Kayne. In particular, I claim that a linearization statement ⟨α, β⟩ is only interpretable if α and β are distinct from each other. In some languages, such as English, nodes are typically nondistinct if they have the same label (though we will see that in many languages, distinctness is more difficult to define than this). Consequently, any phase in which two DPs, for example, must be linearized with respect to each other yields a linearization statement ⟨DP, DP⟩, which causes the derivation to crash. I will claim that parts of classic Case theory can be made to follow from this general principle, which also covers facts that have nothing to do with DPs.

The second proposal, "Beyond Strength and Weakness," is an attempt to predict, for any given language, whether that language will exhibit overt *wh*-movement or not. The claim is that all languages are required to minimize the number of prosodic boundaries of a certain type between *wh*-phrases and the complementizers where they take scope. In some languages, this general prosodic requirement can be met by manipulating the prosody directly. In others, the prosody cannot be manipulated in this

way, and the only way to satisfy the general condition is to move the
wh-phrase closer to the complementizer in question. I claim that we can
predict whether a language can leave *wh* in situ or not by investigating
more general properties of its prosody, which have nothing to do with
wh-questions particularly. The proposal is related to recent works investigating the interaction between prosody and syntax (Zubizarreta 1998,
Ishihara 2001, Szendrői 2001, Arregi 2002, Vicente 2005, and the references cited there), some of which develop the idea that syntactic movements can have the purpose of improving the prosodic structure of the
sentence.

   Both of these proposals attempt to deepen our current explanations for
syntactic phenomena. Classic Case theory has a number of empirical suc-
cesses, many of which are captured by the theory of Distinctness developed in chapter 2. To the extent that that theory succeeds, it not only
broadens the empirical coverage of the existing theories, but offers an answer to a question that we often do not ask: Why should languages have
Case at all? In chapter 3, I turn to another typically unasked question:
Why do some languages have overt *wh*-movement, while others do not?
The goal in that chapter is to find the level of description on which languages do not in fact vary; all languages, I claim, obey a universal prosodic condition on *wh*-questions. The apparent variation can be reduced to
previously observed variations in prosodic systems.

2 Distinctness

p. 71

A number of phenomena in different languages seem to be constrained by a ban on multiple objects of the same type that are too close together. In this chapter I attempt to develop a general theory of bans of this type, which involves (among other things) the relevant formal definitions of "same type" and "too close together." The result is a theory of syntactic phenomena not unlike those handled by the OCP in phonology.[1] We will see that some of the facts that have traditionally been handled by Case theory can be made to follow from more general conditions on syntax.

English multiple sluicing offers a relevant example. English sluicing may in principle involve multiple remnants:

(1) a. I know everyone danced with someone, but I don't know [who] [with whom]
    b. I know every man danced with a woman, but I don't know [which man] [with which woman]

However, multiple sluicing is impossible if the sluicing remnants are DPs:

(2) a. *I know everyone insulted someone, but I don't know [who] [whom]
    b. *I know every man insulted a woman, but I don't know [which man] [which woman]

To handle these facts and others like them, I will propose a new well-formedness condition on the linearization statements used by Kayne's (1994) LCA. The new restriction on ordering statements will have the effect of making multiple syntactic nodes of the same kind impossible to linearize if they are close together in the structure, in a sense to be made precise. Such unlinearizable structures are therefore banned, and are avoided in a number of ways, as we will see.

Much of the work of this chapter will be to formalize notions like "close together" in ways that allow us to cope with the many apparent counterexamples that the theory will face. For instance, we need to be able to distinguish between examples like (2), in which multiple DPs cannot be adjacent to each other, and examples like (3), in which they clearly can:

(3) I gave John a book.

This particular type of example will be discussed in section 2.4.4.2. More generally, I will argue for a theory of facts like the ones in (1)–(3) that makes use of our existing understanding of syntax. The well-formedness of (3) shows us that a theory that simply bans sequences of adjacent DPs is too restrictive; I will try to show in what follows that a theory that makes particular reference to the construction in (1) and (2) misses a generalization, given the widespread appearance of phenomena of this general kind.

I will assume, with Chomsky (1995, 2000, 2001), that the trees generated by the syntax do not contain complete information about linear order, and that fully linearizing the nodes of the tree is one of the tasks performed by the operation of Spell-Out. I will assume that linearization is accomplished via a version of Kayne's (1994) LCA. Spell-Out considers the set A of pairs of asymmetrically c-commanding XPs and X°s in the tree that the syntax gives it, and generates from this a set of instructions for linearization; if $\langle \alpha, \beta \rangle$ is in A, then the *image* of α (that is, the terminals dominated by α) is ordered with respect to the image of β.

I will also assume, again following Chomsky (2000, 2001), that Spell-Out can occur several times in the course of a syntactic derivation; in particular, that it occurs as soon as a *strong phase* has been constructed. Strong phases will include CP, transitive vP,[2] PP, and KP; I will not try to develop a principled theory of the distribution of phase boundaries, leaving this task for future work. Spell-Out sends material from a strong phase to PF, making all of it inaccessible to further syntactic operations, apart from its edge. Following Chomsky (2000, 2001), the edge of a phase will be the highest head of the phase along with its specifiers; in other words, the complement of the phase head will be spelled out to PF. I will crucially assume, following Nissenbaum (2000), that the edge is linearized with the material in the higher phase. I will refer to the material that is sent to PF via Spell-Out as the "Spell-Out domain."

My new proposal will have to do with the content of linearization statements. Consider the linearization of a tree like the one in (4):

(4)

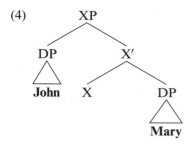

The two DPs in this tree are in an asymmetric c-command relation, and can therefore, following Kayne, be related by a linearization statement. What form should this statement take? In the tree in (4), the two nodes in question are both labeled "DP"; we might therefore take the relevant linearization statement to be ⟨DP, DP⟩. Of course, this ordered pair does not look very helpful for linearization; depending on how these statements are interpreted, it seems to say that a DP precedes itself, or perhaps that one DP precedes the other (without specifying which comes first). We could avoid this result by assigning the two DPs indices, or allowing the linearization statement to refer to the contents of the two DPs (⟨DP(*John*), DP(*Mary*)⟩), or to their positions (⟨DP in specifier of XP, DP complement of X⟩), more or less as we would when informally referring to nodes on the tree.

I want to suggest, however, that these richer ways of referring to positions in the tree are not in fact available to the linearization process. The linearization statement associated with a tree like (4) is indeed ⟨DP, DP⟩. Such statements are uninterpretable, perhaps because the linearization algorithm regards them as self-contradictory instructions to make nodes precede themselves. As a result, any tree that generates linearization statements of this kind cannot be linearized, and the derivation for such a tree crashes at Spell-Out. The upshot of this is that when a subtree is spelled out, if any pair of nodes in that tree cannot be distinguished from each other and are in an asymmetric c-command relation, the derivation crashes. The new condition on linearization is given in (5):

(5) *Distinctness*
    If a linearization statement ⟨α, α⟩ is generated, the derivation crashes.

This condition rejects trees in which two nodes that are both of type α are to be linearized in the same Spell-Out domain, and are in an asymmetric c-command relation (so that a linearization statement relating them is generated).

Which nodes are "of the same type" in the relevant sense? For English and several other languages, it will turn out to be nearly sufficient to say that nodes cannot be distinguished from each other by linearization statements if they have the same label. The tree in (4), for example, is an unacceptable candidate for Spell-Out in English, since the two DPs in it cannot be distinguished from each other, and one c-commands the other.

We will see in section 2.3, however, that this is not the whole story. This is welcome news, since it is not clear that node labels ever have the importance attributed to them above (for one argument that they never do, see Collins 2002). As the chapter progresses, we will find two general classes of cases in which nodes with the same label can be linearized in apparent violation of Distinctness. First, we will see that languages vary in the extent to which they can draw distinctions between projections with the same label. In the particular case of DPs, for example, some languages (like English) treat all DPs as identical, while for others, DPs are the same only if they have identical values for case, grammatical gender, and/or animacy. For most of what follows I will concentrate on languages of the English type, so I will continue to write as though nodes are distinguished simply by their labels, but it is worth bearing in mind that this is not strictly true. Second, lexical projections seem to be very generally immune to Distinctness; the effects we will be concerned with here appear to arise only for interactions between functional projections (this will be particularly important in section 2.4.2).

I believe that the distribution of Distinctness effects may be usefully linked to a distinction standardly drawn in the framework of Distributed Morphology (see Halle and Marantz 1993, Marantz 1997, Embick and Noyer 2006, and the references cited there) between lexical and functional heads (in the Distributed Morphology literature, these are known variously as Roots and Abstract Morphemes, or as L-morphemes and F-morphemes). In much work in the Distributed Morphology framework, these heads enter the syntactic derivation in quite different ways.[3] Lexical heads are Merged as complete lexical items. Functional heads, by contrast, are Merged simply as bundles of features; these bundles of features are subsequently associated with phonological information via a postsyntactic process of vocabulary insertion. To use the terminology of Distributed Morphology, functional heads undergo Late Insertion, while lexical heads undergo Early Insertion.

Suppose that linearization takes place prior to late vocabulary insertion. Several conclusions follow about the kind of information to which linearization statements may be sensitive. When lexical projections are linearized, the lexical items have already been inserted, and linearization may therefore make reference to a rich array of properties distinguishing the heads from each other.[4] When functional projections are linearized, by contrast, vocabulary insertion has not yet taken place, and information that might later serve to distinguish different functional heads is not yet present. As a consequence, Distinctness effects will arise just in interactions between functional heads. Moreover, languages vary in the richness of the feature bundles to which the linearization process for functional heads may make reference. For some languages, prior to vocabulary insertion, a head like D is represented only by the feature [D]; for others, linearization may apparently refer to a richer bundle of features, including features like gender and case.

If we are right in ordering linearization before the full set of phonological features have been inserted for functional heads, we expect these phonological features to have no effect on linearization. For example, it should not matter, as far as linearization is concerned, whether a functional head is overt or not; if a particular vocabulary item is associated with no phonological information at all, it will still need to be linearized, on this view. We will see that this prediction is borne out, for instance, in section 2.2.1.1. The one exception to this generalization that we might expect to find would involve instances in which the syntax itself was responsible for something being phonologically null. Copies left behind by movement, for example, will not be linearized, perhaps because their phonologically null status is predictable even before vocabulary insertion takes place.

The proposal given here makes use of Kayne's (1994) idea that some conditions on syntactic trees may follow from the need to render those strings as linearly ordered strings of words. The chapter will not need to adopt all of the particulars of Kayne's (1994) proposal, however. For example, Kayne's idea that the specifier of XP is not fully dominated by XP will play no role in the account. Similarly, I will remain agnostic about whether a linearization statement $\langle \alpha, \beta \rangle$ is invariably interpreted as "α precedes β," allowing for the possibility that heads, for instance, can be specified as occurring to the right of their complements.

The chapter proceeds as follows. In section 2.1, I review a number of cases of Distinctness. Section 2.2 argues that the relevant condition should be a condition on linearization; I discuss more fully the mechanics of linearization, and we will see that Distinctness effects arise just when two nodes of the same type are linearized in the same Spell-Out domain. In section 2.3, I consider the set of types of nodes to which linearization statements may refer; we will see that these types vary from language to language. Section 2.4 outlines some of the ways Distinctness violations are circumvented, including the removal of offending structure, addition of structure that insulates the potentially offending nodes from each other by adding a phase boundary between them, the suppression of movement operations that would otherwise create Distinctness violations, and the triggering of movement operations that separate potentially offending nodes. Section 2.5 considers the extent to which classic Case theory may be replaced by the theory developed here. Section 2.6 concludes the chapter.

2.1   Distinctness Violations

In this section I consider a number of instances in which linearization fails because the objects to be linearized in a single Spell-Out domain are insufficiently distinct. These cases are simply intended to illustrate the general idea; the exact structures in these examples are often unclear, and are largely irrelevant for our purposes. I begin looking at more clearly understood examples in section 2.2. (p.16)

2.1.1   Multiple Sluicing, Multiple Exceptives, *even*

As pointed out by Sauerland (1995) and Moltmann (1995), English exhibits an odd constraint on several types of constructions involving multiple ellipsis remnants. Multiple remnants with exceptives, ellipsis with *even*, and sluicing are all in principle possible in English:

(6) a. Every man danced with every woman, except [John] [with Mary]
    b. Every man danced with every woman, even [John] [with Mary]
    c. I know everyone danced with someone, but I don't know [who] [with whom]

Such constructions are impossible in English if the remnants are both DPs:[5]

(7) a. *Every man admired every woman, except [John] [Mary]
    b. *Every man admired every woman, even [John] [Mary]
    c. *I know everyone insulted someone, but I don't know [who] [whom]

If we assume that the ellipsis remnants are both in the same Spell-Out domain (perhaps because a movement operation has moved both of them out of the domain of ellipsis), then these facts follow from Distinctness;

in (7), the two DPs cannot be linearized. I return to these facts in section 2.3.2, where we will see that they are a point of crosslinguistic variation.

2.1.2 DP-Internal Arguments

Within a gerund, the subject and object of the verb on which the gerund is based may optionally be expressed as PPs headed by *of*:

(8) a. the singing [of the children]
   b. the singing [of songs]

However, only one argument may surface in this way. If both arguments are to be expressed, they must be introduced with different prepositions:

(9) a. *the singing [of songs] [of the children]
   b. the singing [of songs] [by the children]

Bans on multiple NP-internal arguments with identical morphology are crosslinguistically common, as Alexiadou (2001) as well as Alexiadou, Haegeman, and Stavrou (2007) observe (both the examples below are from Alexiadou, Haegeman, and Stavrou 2007, 543):

(10) a. *i   silipsi   tu Jani   tis astinomias (Greek)
        the capture the John-GEN the police-GEN
        'the capture of John by the police'
    b. *l'afusellament de l'escamot d'en Ferrer Guardia (Catalan)
        the-execution of the-squad of   Ferrer Guardia
        'the squad's execution of Ferrer Guardia'

The ban on nominalization of verbs with double objects (Kayne 1984; Pesetsky 1995) might be related to these facts:

(11) a. *the gift [of John] [of a book]
    b. the gift [of a book] [to John]

For the contrasts in (9) and (11) to follow from Distinctness, there will have to be some relevant difference between PPs headed by *of*, on the one hand, and those headed by *to* and *by*, on the other. One possibility would be to view *of* in these constructions as a pronounced K(ase) head, while *to* and *by* are actually Ps, so that the prepositional phrases in (9b) and (11b) are a KP and a PP, respectively.

If we say nothing else, then the well-formedness of (12) is puzzling, since it appears to contain two PPs, one headed by *to* and the other by *by*:

(12) the gift [of a book] [to John] [by Mary]

Either we need some further difference between PPs headed by *to* and ones headed by *by*, such that the two PPs may be successfully linearized, or we must posit a phase boundary within the structure of NP, which will keep these two PPs safely segregated. I have no evidence bearing on the question, so will leave it here for now.

2.1.3 Causatives

A number of languages have a causative construction in which the morphological form of the causee varies depending on the transitivity of the caused predicate. When the caused predicate is intransitive, the causee is marked like an object, but when the caused predicate is transitive, the causee is marked like an indirect object:

(13) a. Jean a   fait   manger **Paul** (French)
        Jean has made eat-INF Paul (Kayne 2004, 193)
        'Jean made Paul eat'
     b. Jean a   fait   manger la   tarte **à Paul**
        Jean has made eat-INF the pie   to Paul
        'Jean made Paul eat the pie'

(14) a. Elena fa   lavorare **Gianni** (Italian)
        Elena makes work-INF Gianni (Guasti 1997, 126)
        'Elena makes Gianni work'
     b. Elena fa   riparare la   macchina **a Gianni**
        Elena makes repair-INF the car   to Gianni
        'Elena makes Gianni repair the car'

(15) a. Mehmet **Hasan-ı**   öl-dür-dü (Turkish)
        Mehmet Hasan-ACC die-CAUS-PAST (Aissen 1979, 8)
        'Mehmet caused Hasan to die'
     b. Hasan **kasab-a**   et-i   kes-tir-di
        Hasan butcher-DAT meat-ACC cut-CAUS-PAST
        'Hasan had the butcher cut the meat'

An extensive literature on this type of causative (see Kayne 1975, 2004; Rouveret and Vergnaud 1980; Burzio 1986; Davies and Rosen 1988; Guasti 1993, 1997; den Dikken 2006b; and the references cited there) establishes that these examples are effectively monoclausal; in particular, the infinitival verb is apparently stripped of its functional structure. For instance, passive morphology on the causative verb affects the logical arguments of the caused predicate:

(16) a. Gianni è stato fatto lavorare
Gianni is been made work-INF
a lungo (Italian)
for.a.while (Guasti 1997, 128, 130)
'Gianni has been made to work for a while'
b. La macchina è stata fatta riparare a Gianni
the car   is been made repair-INF to Gianni
'Gianni has been made to repair the car'

Thus, the transitivity of the entire clause appears to be determined by the voice of the causative verb. In other words, there appears to be only one instance of *v*P in the clause, and it is the *v*P that introduces the causer.[6]

If this is correct, then we are led to a structure for this type of causative in which the arguments of the caused predicate are all in a single Spell-Out domain; there are no embedded instances of *v*P to introduce phase boundaries between these DPs (see den Dikken 2006b, in particular, for a proposed structure of this type). Distinctness therefore demands that the DPs be kept separate from each other. When the embedded predicate is intransitive, no difficulty for linearization arises. When the embedded predicate is transitive, however, the causee and the object of the caused verb cannot both be ordinary DPs. Consequently, it is just in cases of causativized transitives that the causee must be made into a PP or KP. Suppose that PP and KP are phases (on the phasal status of PP, see Abels 2003). Embedding the causee in a PP or KP will then put the causee into its own separate Spell-Out domain, safely insulating it from the direct object of the causativized verb.

2.1.4 DP Predication in Predicate-Initial Languages

In Tagalog we find another instance of an apparent ban on structurally adjacent DPs, which can be attributed to Distinctness. Tagalog is a predicate-initial language, which in principle allows any kind of phrase to be a predicate. The examples in (17) show the behavior of VP, AP, PP, and NP predicates, respectively:

(17) a. Umuwi si Juan (Tagalog)
went.home Juan
'Juan went home'
b. Mataas si Juan
tall   Juan
'Juan is tall'

   c. Tungkol sa balarila   ang libro
      about   grammar the book
      'The book is about grammar'
   d. Guro   si Maria
      teacher   Maria
      'Maria is a teacher'

Tagalog also allows DPs to be predicates. However, when a DP is a predicate, the regular predicate-initial word order is blocked; the subject must be initial:

(18) a. Si Gloria ang pangulo. (Tagalog)
         Gloria the president
         'Gloria is the president'
     b. *Ang pangulo si Gloria

We find similar facts in Irish, another predicate-initial language (Carnie 1995, 18, 20):

(19) a. Leanann   an t-ainmní an briathar i nGaeilge (Irish)
         follow.PRES the subject   the verb   in Irish
         'The subject follows the verb in Irish'
     b. Is platapas (é)   Seán
         C platypus AGR John
         'John is a platypus'

Again, despite the fact that Irish predicates are generally initial, DP predicates must follow the subject (Carnie 1995, 20):

(20) Is é   Seán an platapas (Irish)
      C AGR John the platypus
      'John is the platypus'

    Distinctness might allow us to make sense of the fact that, just in the case in which the subject and the predicate have the same label, the ordinary word order in both Tagalog and Irish is reversed. Accounts of the derivation of predicate-initial word order vary, but may be classified into two major types. In one type of account, the predicate head, or some projection of that head, moves to a position above the subject (Lee 2000, Massam 2000, and Rackowski and Travis 2000). In another, the predicate is left in its usual place, but the subject is lowered into the projection of the predicate (Chung 1990 and Sabbagh 2008):

(21) a.

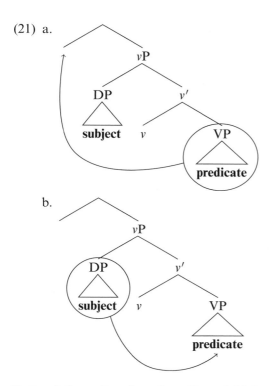

b.

Both of these theories of predicate-initial word order have the consequence that the subject and the predicate are structurally quite close together. Consequently, we would expect Distinctness to ban the movement depicted here just in cases in which the subject and the predicate are both DPs. Indeed, this appears to be what we find.

2.1.5 English Quotative Inversion, Locative Inversion; French Stylistic Inversion

208 n 10

Consider the facts in (22):

(22) a. "It's cold," said John
 b. "It's cold," said John to Mary
 c. *"It's cold," told John Mary

Here the subject is in some postverbal position, and the verb has apparently raised past it, as has the quote (Collins and Branigan 1997). We find a similar pattern of facts with Locative Inversion (Bresnan and Kanerva 1989, Doggett 2004, Wu 2008, and the references cited there):

(23) a.   [Into the room] walked a man
     b.   [Into the room] walked a man in the afternoon
     c.   *[Into the room] kicked a man a ball

Accounting for the data in (23) will require our first departure from standard assumptions about phases. In order for the two DPs in sentences like (23c) to be in the same phase, I will need to assume that the base position of the subject is not, in fact, the highest position in the vP phase. I will retain the name vP for the projection in which the external subject is base-generated, and add another functional projection above this one, which I will call v_CP, projected by the phase head v_C:

(24)

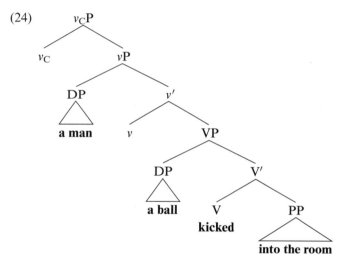

In the tree in (24), v_C is related to v in the way that C is related to T. In particular, just as T inherits its ϕ-features from the phase head C (Chomsky 2005), v inherits its ability to Agree with and license objects from the phase head v_C. Thus, v_C is responsible for making v transitive, and is absent when v is intransitive. We will see further evidence for the existence of v_C in section 4.4.2.3, where we will find it realized as overt morphology in Kinande.

Because v_C is a phase head, any phrase that is to exit the lowest phase will have to move to its edge. In the locative inversion example in (23c), for example, the locative PP moves to a specifier of v_CP. Note, too, that in order to get the correct word order, we must assume that the English verb moves at least to v_C, as in (25).

(25)

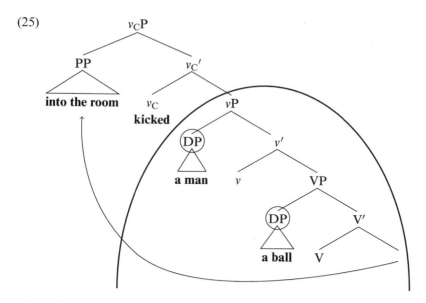

After the PP moves to the specifier of v_CP, the phase head v_C will trigger Spell-Out of its complement to PF. Let us consider how this Spell-Out domain is to be linearized. Many of its terminal nodes are occupied by copies in movement chains; the PP has moved out of the Spell-Out domain, as has the verb *tell*. All that is left to linearize in this part of the tree are the two DPs *a man* and *a ball*. By hypothesis, the linearization process has no way of distinguishing between these two instances of DP. As a result, linearization of this material generates the ordering statement $\langle$DP, DP$\rangle$, which is self-contradictory and causes the derivation to crash.[7]

As we will continue to see below, the presence of an offending linearization statement is always enough to make the derivation crash. Crucially, the derivation crashes even if the structure is in fact linearizable without using the offending statement. In the particular case of (10), for example, if we disregard Kayne's suggestion that maximal projections do not have their specifiers in their images, we might expect to be able to linearize the tree without using the $\langle$DP, DP$\rangle$ pair. The higher DP, after all, c-commands VP, and we might be able to conclude from that that this DP precedes everything in VP, including the lower DP. Apparently this is not enough to save the structure; the system gives up as soon as an uninterpretable linearization statement appears.

French has a phenomenon known as Stylistic Inversion (see Kayne 1972, Kayne and Pollock 1978, Déprez 1988, Valois and Dupuis 1992, and the references cited there for discussion), which is subject to conditions that resemble those on Quotative Inversion and Locative Inversion in English. French subjects may be postposed in *wh*-extraction contexts (Kayne and Pollock 1978, 595):

(26) a. Quand partira   ton   ami? (French)
        when   will.leave your friend
        'When will your friend leave?'
     b. Je me demande quand partira   ton   ami
        I me ask   when will.leave your friend
        'I wonder when your friend will leave'

However, Stylistic Inversion is impossible when it would yield a sentence with two DP arguments following the verb (Valois and Dupuis 1992, 327; Collins and Branigan 1997, 17):

(27) a. *Je me demande quand mangera sa   pomme Marie (French)
         I me ask   when will.eat her apple   Marie
         'I wonder when Marie will eat her apple'
     b. *Je me demande quand mangera Marie sa   pomme
         I me ask   when will.eat Marie her apple
         'I wonder when Marie will eat her apple'

(28) Quel livre a   donné Marie à Paul?
     which book has given Marie to Paul
     'Which book has Marie given to Paul?'

In (27), linearization fails; as in the English inversion cases, the linearization process for the postverbal domain includes a statement $\langle DP, DP \rangle$, which will cause the derivation to crash. In (28), we can see that the ban is not simply on multiple postverbal phrases; a DP may appear together with a PP in the postverbal field. For insightful discussion of a number of phenomena of this type, see Alexiadou and Anagnostopoulou 2001, 2007.

2.2 The Mechanics of Distinctness

The account I have sketched of phenomena like those in the previous section is sensitive to syntactic structure in a way that I have not yet shown to be necessary. At this stage, we might wonder whether so much syntactic heavy machinery is really called for to deal with the phenomena under discussion. Instead, one might posit a ban on "stuttering" that penalized linearly adjacent words of the same kind.[8] We might claim, in other words, that the phenomena I have just discussed have nothing to do with syntax.

In this section, I will try to defend the claim that Distinctness effects are crucially sensitive to syntactic structure and are not about linear adjacency. The syntactic conditions that give rise to Distinctness effects, I will argue, often do result in linear adjacency between the syntactic objects involved, but we will see that adjacency is neither necessary nor sufficient for Distinctness effects to arise. In section 2.2.1, I will show that linear adjacency is insufficient to get Distinctness effects; section 2.2.2 will be devoted to showing that it is unnecessary.

2.2.1 Linear Adjacency without Distinctness

The case studies in this section will involve phenomena that Distinctness can handle. We will also see, however, that the objects regulated by Distinctness are sensitive, not to linear adjacency, but to Spell-Out domains.

2.2.1.1 Perception/Causation Verb Passives, Doubl-*ing*, Italian Double-Infinitive Filter
A number of phenomena involve bans on sequences of adjacent verbs, which can be treated in terms of Distinctness. We will see that these bans are sensitive to phase boundaries.

Sentences like those in (29) cannot be passivized (Fabb 1984; Santorini and Heycock 1988):

(29) a. We saw John leave
     b. We let John leave
     c. We made John leave

(30) a. *John was seen __ leave
     b. *John was let __ leave
     c. *John was made __ leave

We can rule out (30) with Distinctness, assuming a tree for the highest *v*P in these examples something like (31):

(31)

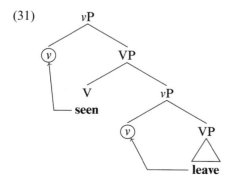

(31) cannot be linearized; there are two heads v, and so there will be a linearization statement of the form $\langle v, v \rangle$, which will make the linearization process crash. Note that I continue to assume, as I did in section 2.1.5, that the English verb raises at least to v.

Why should there be a distinction between the examples in (29), in which the matrix verb is active, and the examples in (30), in which it is passive? Recall that we are following Chomsky (2000, 2001) in assuming that the extended vP projection is a strong phase just in case the verb is transitive. In section 2.1.5, I posited a phase head v_C, which appears only when the verb is transitive and supplies v with the features it needs to license a direct object, via feature inheritance (Chomsky 2005). The subtree in (31) contains no phase heads, since both verbs are intransitive; the lower verb is presumably unaccusative, and the higher verb is passive, and therefore no instances of v_C appear. By contrast, in the corresponding subtree for the examples in (29), the higher v is transitive, and hence the phase head v_C must be Merged to supply the features v needs to license the object. Since v_C is a phase head, its complement will undergo Spell-Out; as before, the verbs will raise to the highest vP projection:

(32)

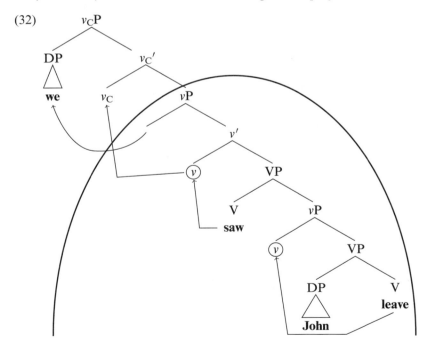

In (32), the phase head v_C donates to v the features necessary to license the object. Because v_C is a phase head, it also introduces a Spell-Out boundary, which makes it possible to separate the two instances of *v* from each other. Both verbs undergo head movement to the highest *v*P projection, and in the case of the higher verb *saw*, this means that the higher *v* will be carried out of the lower Spell-Out domain, protecting it from linearization with the lower instance of *v*.

The theory developed here, then, is one in which the relevant difference between (29) and (30) has to do with the transitivity of the higher verb. Of course, there are other differences between the examples in (29) and (30)—for instance, the two verbs are string-adjacent in (30) but not in (29). But the relevant notion clearly is not simply string adjacency, since *wh*-traces, unlike NP-traces, do relevantly intervene between the verbs:

(33) a. [How many prisoners] did you see ___ leave?
    b. [How many prisoners] did you let ___ leave?
    c. [How many prisoners] did you make ___ leave?

In terms of the theory developed here, this contrast between *wh*-traces and NP-traces follows from the distribution of strong phases. The higher verb in the examples in (33) is associated with the apparatus of case assignment to the *wh*-moved NP. The higher instance of *v*P is dominated by a v_CP, which renders *v* transitive and also introduces a Spell-Out domain that can separate the two instances of *v*. Head movement moves one *v* across the Spell-Out boundary, and as a result, linearization succeeds.[9] The *v*P of the higher clause in (30), by contrast, is intransitive, and the two instances of *v* must therefore undergo Spell-Out in the same domain. The trees for (30a) and (33a) are given in (34):

(34) a.

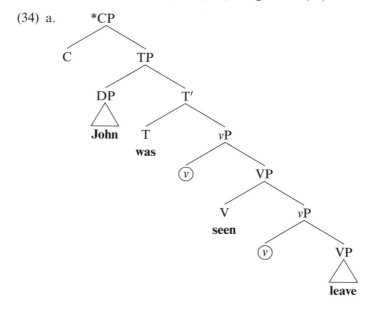

b.

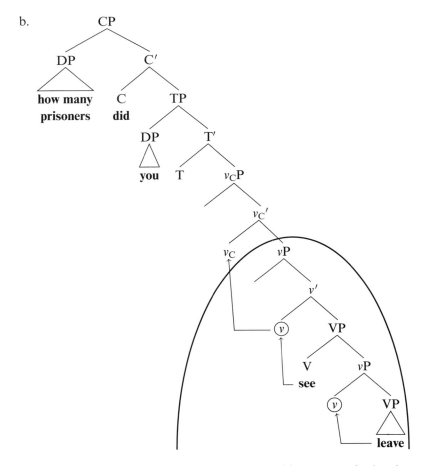

The relevant difference between the two trees, on this account, is that in (34b) the higher *v* is transitive, assigning a θ-role to *you* and Case to *how many prisoners*. Since this instance of *v* is transitive, it is dominated by a projection of the phase head v_C, which introduces a Spell-Out domain. Head movement of each verb to the highest projection of *v* separates the two instances of *v* from each other, and the tree is therefore linearizable. In (34a), by contrast, the higher instance of *v* is intransitive, and hence not associated with a phase head; as a result, both instances of *v* are spelled out in the same domain, and the attempt to linearize that Spell-Out domain will crash, since one of the linearization statements will be $\langle v, v \rangle$.

At least two potential worries are worth addressing at this point. First, the two instances of *v* that are putatively responsible for the Distinctness violation are phonologically null. Why do they need to be linearized at all, given that they will not be pronounced? And second, the discussion of (34) has centered entirely on the status of the higher *v*P; when this is intransitive, as in (34a), the result is ill-formed, and when it is transitive, as in (34b), the result is well formed. What about the transitivity of the lower *v*P?

The answer to the first question is related to a proposal made in the introduction to this chapter. I suggested there that linearization takes place before insertion of functional heads, but after insertion of lexical heads (following a claim from Distributed Morphology that insertion of functional heads takes place postsyntactically). Because linearization takes place before functional heads have undergone vocabulary insertion, the structure to which linearization applies will be one in which functional heads are represented only as feature bundles; in particular, no phonological information has yet been attached to them. This is the answer to the first question above. Because linearization takes place before phonological information has been attached to functional heads, the question of whether a functional head is phonologically null or not should be of no consequence to linearization.

The second question above had to do with the status of embedded *v*P. The contrast in (34) is entirely concerned with the higher *v*P; when this is intransitive, as in (34a), the result is ill-formed, and when it is transitive, as in (34b), the result is well formed. What about the lower *v*P?

In fact, we do not expect the lower *v*P to affect the grammaticality of this type of example, given our assumptions about how Spell-Out operates. If we are correct in taking the domain of Spell-Out to be the sister of the phase head, the transitivity of the lower *v*P should make no difference. The examples in (34) both involve an intransitive lower *v*P, but suppose we consider a version of (34a) with a transitive lower *v*P:

(35)   *

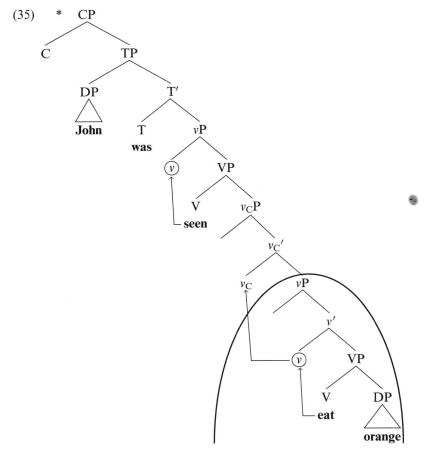

Even if the lower *v* is transitive, the example is correctly ruled out. Head
movement of the V to *v*$_C$ will carry the lower instance of *v* into the Spell-
Out domain containing the higher instance of *v*, yielding an unlinearizable tree. We therefore correctly expect that the transitivity of the lower
v is unimportant for Distinctness; only the transitivity of the higher *v*
should matter.

By similar logic, this account deals with an otherwise mysterious gap in
English subcategorization; although, as we have seen, there are several
transitive verbs that take bare infinitive complements, there seem to be
no intransitive verbs that do:

(36) a.   John seems __ tired
     b.   John seems __ a fine fellow
     c.   *John seems __ enjoy movies

Such verbs would always yield unlinearizable sentences, as in the passive case above; there would be no transitive vP to provide a strong phase boundary to separate the two verbs.

English has another phenomenon reminiscent of the one discussed above, originally discovered by Ross (1972). Ross notes that (37d) is ill-formed, which is surprising in light of the well-formedness of (37a–c):

(37) a. It continued to rain
  b. It continued raining
  c. It's continuing to rain
  d. *It's continuing raining

Ross proposes a "Doubl -*ing* Constraint"; adjacent verbs with the ending -*ing* are banned. The relevant configuration can apparently be broken up by traces of A′-movement, though not by traces of A-movement (or, to put the facts in terms of the theory under development here, the higher v can be shielded from the lower v by a phase boundary, just when the higher v is transitive). Example (37d) shows that A-movement traces can-not relevantly intervene, while (38) shows that *wh*-traces can:

(38) the children [that I was watching __ playing]

Ross demonstrates in a number of ways that his constraint cannot simply be a ban on string-adjacent verbs ending in -*ing*. For instance, he notes the well-formedness of examples like the ones in (39):

(39) a. I watched [a man who had been fly**ing**] describ**ing** it to some chicks.
  b. His expect**ing** [breath**ing** deeply to benefit us] is hopelessly naive.
  c. Waldo keeps molest**ing** [sleep**ing** gorillas].

Ross (1972, 74) concludes that his constraint only applies to verbs that "were in immediately adjacent clauses in remote structure." In our terms, we can describe the condition in terms of Spell-Out domains. In (39a), for example, the two instances of -*ing* are in separate clauses (and in any case are not in a c-command relation), and will thus never need to be linear-ized together. We can also account for the facts in (39b,c), as long as the functional head occupied by -*ing* is at or higher than v_C, and as long as English objects remain inside vP. In the subtree in (40), I have put -*ing* in an Asp(ect) head just above v_CP:

(40)

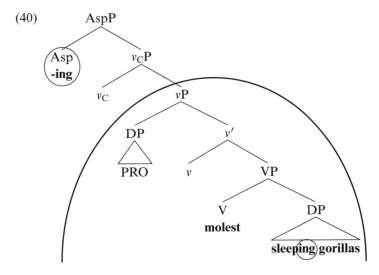

In (40), the Asp head *-ing* is protected from the *-ing* of *sleeping*; v_C induces Spell-Out of *sleeping*, but not of the higher Asp head.[10] The DP subject presumably later raises to the external subject position (not shown), thus separating the two DPs with a Spell-Out boundary.

Finally, Longobardi (1980) discusses a phenomenon in Italian that shares many of the properties of the English facts discussed above. Under certain circumstances, he notes, sequences of infinitives are unacceptable in Italian:

(41) *Paolo potrebbe sembrare __ dormire tranquillamente (Italian)
     Paolo could   seem-INF   sleep-INF quietly
     'Paolo could seem to sleep quietly'

In (41), the verbs *sembrare* and *dormire* are in the same Spell-Out domain. The *v*P associated with *sembrare* is not a Case-assigning one, and hence does not create a strong phase boundary. As in the English case discussed above, a trace of A′-movement can block the relation between the two verbs:

(42) Ecco l'uomo [che puoi   vedere __ portare ogni giorno dei
     here's the-man that you.can see-INF take-INF every day   some
     fiori   a Mario]
     flowers to Mario
     'Here's the man that you can see take some flowers to Mario every
     day'

Again, we can capture this fact in terms of phase boundaries. The verb
vedere 'see-INF' in (42) is associated with a transitive *v*P, which checks
Case on the relative operator. This strong phase protects the two instances of the infinitival suffix from each other. I return to this phenomenon in section 2.4.2.1, where we will see that violations of the
double-infinitive filter may be rescued by restructuring.

The double-infinitive filter is certainly not universal; it is not found in
French or Spanish, for instance, or in English, for that matter:

(43) a. Paul pourrait sembler dormir   calmement (French)
        Paul could   seem-INF sleep-INF calmly
     b. Pablo pudo parecer   dormir   tranquilamente (Spanish)
        Pablo could seem-INF sleep-INF quietly
     c. Paul is likely to seem to sleep quietly

I do not know why Italian differs from French, Spanish, and English in
this way. The most straightforward way to save the examples in (43)
would be to declare that raising infinitives in these languages are phases
(or at least are capable of being phases), so for the time being, this is
what I will say.[11]

These three phenomena—the English ban on passives of verbs taking
bare infinitive complements, the English doubl-*ing* filter, and Italian's
double-infinitive filter—clearly ought to be accounted for within the same
theory. They share a number of intriguing properties, particularly sensitivity to phase boundaries (previously described as sensitivity to the presence of Case-marked traces). In the account developed here, all of these
phenomena are instances of the effects of Distinctness; the functional
heads associated with the verbs cannot be linearized if they are both in
the same Spell-Out domain.

2.2.1.2 Differential Object Marking Chaha, Spanish, Hindi, and Miskitu
all have a case particle with an intriguing distribution. I will argue that
the particle appears to distinguish DPs that would otherwise all have to
be linearized in the same Spell-Out domain. We will see, again, that phase
boundaries are crucial, and that string adjacency is not relevant.

Chaha (a Semitic language of Ethiopia) exhibits a form of object shift,
obligatorily moving specific objects to a position higher than the nonspecific ones:[12]

(44) a. Č'am^{w}ɨt nɨmam <u>ambɨr</u>   tɨčəkɨr (Chaha)
        Č'.   normally cabbage cooks
        'Č'am^{w}ɨt normally cooks cabbage'

  b. *Č'amʷɨt <u>ambɨr</u>   nɨmam   tɨčəkɨr
    Č'.   cabbage normally cooks

(45) a. *Č'amʷɨt nɨmam   <u>ambɨr</u>   xʷɨta tɨčəkʷɨnn
    Č'.   normally cabbage the   cooks
  b. Č'amʷɨt <u>ambɨr</u>   xʷɨta nɨmam   tɨčəkʷɨnn
    Č'.   cabbage the   normally cooks
    'Č'amʷɨt normally cooks the cabbage'

When an SOV sentence contains two animate DPs[13] and the object is specific, the object is obligatorily marked with a prefix *yə-*:

(46) a. Gɨyə **yə**-fərəz nəkʷəsənɨm (Chaha)
    dog **yə** horse bit
    'A dog bit a (specific) horse'
  b. Gɨyə fərəz nəkəsəm
    dog   horse bit
    'A dog bit a (nonspecific) horse'

We can analyze these facts in terms of Distinctness. The account will involve making some guesses about Chaha's (rather sketchily explored) syntax, which I will try to be as explicit about as possible. The general idea will be that *yə-* appears to distinguish between DPs that are too close together to linearize.

Let us suppose, first of all, that the subject always raises to Spec TP, and that object shift involves overt movement to a specifier of v_CP (that is, to the edge of the v_CP phase). There is some reason to think that the verb always c-commands the subject; negation on the verb licenses NPIs in both subject and object position, for instance:

(47) a. Namaga attɨkar   ansɨyə (Chaha)
    Namaga anything NEG-bought
    'Namaga didn't buy anything'
  b. *Namaga attɨkar   sɨyəm
    Namaga anything bought

(48) a. attɨsəβ bɨk'ʷrə ansɨyə
    anyone mule   NEG-bought
    'No one bought a mule'
  b. *attɨsəβ bɨk'ʷrə sɨyəm
    anyone mule   bought

Thus, I will depict the verb as having raised to a C that takes its complement on the left; I will leave aside the problem of how to reconcile this with the LCA, having no way at the moment to choose among the various technical options available.[14]

Let us consider, first of all, the example in (46b), in which object shift has not taken place:

(49)

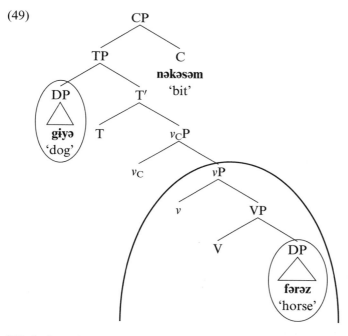

(49) is linearizable. The only repeated types of nodes are the two DPs, and these are spelled out in different Spell-Out domains; the v_CP phase spells out the domain that is boxed in the tree in (49), which contains only one of the two DPs.

Next consider (46a), where object shift to Spec v_CP has taken place. Here the object must be marked with *yǝ-*, which I will analyze as heading a KP dominating DP:

(50)

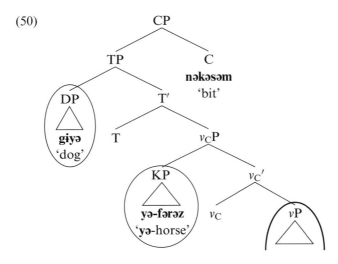

Because the object has undergone object shift, it has moved to the edge of the strong phase v_CP and is therefore linearized with the higher phase. I propose that (50) obeys Distinctness only because of the KP dominating the lower DP. If we are right to assume (as we did in section 2.1.3, for example) that KP is a phase, then the DP dominated by this KP is spelled out in the KP phase, and is therefore safely linearized by the time the higher DP is introduced.[15]

Additional evidence for the approach to *yə-* developed here comes from the behavior of sentences with more than two arguments. Here the indirect object must always be marked with *yə-*, regardless of specificity:

(51) Č'am^wɨt y-at   mɨs fɨraŋk awəčnɨm (Chaha)
     Č'am^wɨt **yə**-one man money gave
     'Č'am^wɨt gave money to a (specific or nonspecific) man'

Thus, *yə-* is not simply a marker for specific direct objects. These facts follow from the account. In (51), there is no way to avoid having the indirect object in the same Spell-Out domain with another DP; if the indirect object shifts, it is in the higher Spell-Out domain with the subject, and if it does not it is in the lower Spell-Out domain with the object. The *yə-* prefix is therefore obligatory, regardless of specificity. Informally, *yə-* acts as a "spacer," which appears whenever DPs are too close together. As we should expect on this account, sentences with multiple internal arguments marked with *yə-* are impossible:

(52) a. Č'amʷɨt yə-tkə   xʷɨta gɨyə awəčnɨm (Chaha)
         Č'amʷɨt yə-child the   dog gave
         'Č'amʷɨt gave the child a/the dog'
     b. *Č'amʷɨt yə-tkə   xʷɨta yə-gyə awəčnɨm
         Č'amʷɨt yə-child the   yə-dog gave

The distribution of Chaha *yə-* has parallels in a number of other languages. Hindi, Spanish, and Miskitu are all like Chaha in having a morphological marker that appears in specific animate direct objects[16] and all indirect objects. In Hindi the marker is *-ko*, in Spanish it is *a*, and in Miskitu it is *-ra*:

(53) a. Ravii (ek) gaay kʰariidnaa caahtaa
         Ravi one cow to.buy   wish
         hai (Hindi)
         AUX (Mohanan 1994a, 79, 80, 85)
         'Ravi wishes to buy a (nonspecific) cow'
     b. Ravii ek   gaay-ko kʰariidnaa caahtaa hai
         Ravi one cow-**ko** to.buy   wish   AUX
         'Ravi wishes to buy a (specific) cow'
     c. Ilaa-ne ek   haar *(-ko) utʰaayaa
         Ila-ERG one necklace-**ko** lifted
         'Ila lifted a necklace'
     d. Ilaa-ne mãã-ko   baccaa diyaa
         Ilaa-ERG mother-**ko** child   gave
         'Ila gave a/the child to the mother'

(54) a. Laura escondió un prisionero durante dos
         Laura hid   a prisoner for   two
         años (Spanish)
         years (Torrego 1998, 21, 40)
         'Laura hid a (nonspecific) prisoner for two years'
     b. Laura escondió a un prisionero durante dos años
         Laura hid   *a* a prisoner for   two years
         'Laura hid a (specific) prisoner for two years'
     c. Golpeó   (*a) la mesa
         he/she.hit *a*   the table
         'He/she hit the table'
     d. Describieron   un maestro de Zen al   papa
         they.described a master of Zen *a*-the pope
         'They described a Zen master to the pope'

(55) a. Yang aaras (kum) atkri (Miskitu)
      I   horse a   bought (Ken Hale, p.c.)
      'I bought a horse'
   b. Yang aaras-ra atkri
      I   horse-*ra* bought
      'I bought a/the (specific) horse'
   c. Yang tuktan ai yaptika-ra brihbalri
      I   child his mother-*ra* brought
      'I brought the child to his mother'

Also as in Chaha, these markers normally cannot appear on multiple DPs in a sentence; in ditransitive sentences, the marker must appear on the indirect object and typically does not appear on the direct object (Hindi from Mohanan 1994a, 85; Spanish from Torrego 1998, 133–134; Miskitu from Ken Hale, personal communication):

(56) a. Ilaa-ne mãã-ko   baccaa diyaa (Hindi)
      Ilaa-ERG mother-*ko* child   gave
      'Ila gave a/the child to the mother'
   b. *Ilaa-ne mãã-ko   bacce-ko diyaa
      Ilaa-ERG mother-*ko* child-*ko* gave
      'Ila gave a/the child to the mother'

(57) a. Describieron un maestro de Zen al   papa (Spanish)
      they.described a master of Zen *a*-the pope
      'They described a Zen master to the pope'
   b. *Describieron a un maestro de Zen al   papa
      they.described *a* a master of Zen *a*-the pope

(58) a. Yang tuktan ai yaptika-ra brihbalri (Miskitu)
      I   child his mother-*ra* brought
      'I brought the child to his mother'
   b. *Yang tuktan-ra ai yaptika-ra brihbalri
      I   child-*ra* his mother-*ra* brought

   Torrego attributes to Strozer (1976) the discovery that there is in fact a class of verbs in Spanish that allow both objects of ditransitives to be marked with *a*, somewhat marginally (Torrego 1998, 134):

(59) ?Mostré / presenté al   alumno al   professor (Spanish)
      I.showed I.presented *a*-the student *a*-the teacher
      'I showed/introduced the student to the teacher'

Torrego offers arguments that the direct objects of such verbs are structurally higher than those of verbs like the one in (59). She suggests a general "exclusion of structures that have [two DPs marked with *a*] in the
same Case-checking domain" (Torrego 1998, 134), which is certainly
compatible with the approach developed here; if the direct object in (59)
can raise out of its strong phase into the strong phase occupied by the
subject, then it and the indirect object can be linearized separately.

Rodríguez-Mondoñedo (2007) also discusses the status of examples like
(59) in which both objects are marked with *a*; his starting example is (his
(16), p. 224):

(60) ??Juan le   presentó   a María a Pedro (Spanish)
     John 3.DAT introduced **a** Mary **a** Peter
     'John introduced Mary to Peter'

He notes that examples of this type are not perfectly well formed, but that
they improve when one of the arguments is extracted (as in (61a)) or
heavy (as in (61b)) or if a prosodic break is put between the objects (as
in (61c)):

(61) a. A Pedro, Juan le   presentó   a María (Spanish)
        a Peter   John 3.DAT introduced **a** Mary
        'To Peter, John introduced Mary'
     b. Juan le   presentó   a María de las Nieves a Pedro
        Juan 3.DAT introduced **a** María de las Nieves **a** Pedro
        Vargas Prada
        Vargas Prada
        'John introduced María de las Nieves to Pedro Vargas Prada'
     c. Juan le   presentó   a María, a Pedro
        John 3.DAT introduced **a** Mary   **a** Peter
        'John introduced Mary, to Peter'

As Rodríguez-Mondoñedo notes, these data suggest that multiple *a*-
marked objects must be kept separate from each other somehow; he
develops an account that makes use of Distinctness.

Bhatt and Anagnostopoulou (1996) note that when both the direct object and the indirect object are marked with *-ko* in Hindi, the word order
becomes fixed; the direct object must precede the indirect object[17] (p. 14):

(62) Ram-ne   Bill-ko Lila-ko di-yaa (Hindi)
     Ram-ERG Bill-*ko* Lila-*ko* give-PFV
     'Ram gave Bill to Lila' / * 'Ram gave Lila to Bill'

Again, we can interpret this (as Bhatt and Anagnostopoulou do) as an effect of obligatory object shift; if the direct object and the indirect object are both marked with *ko*, the object must shift out of the phase containing the indirect object.[18]

Let us extend the account of Chaha *yə-* to Hindi *-ko*, Spanish *a*, and Miskitu *-ra* (although the latter languages have word order that is freer than Chaha's in ways that generally make it more difficult to show that the same kinds of movements of DPs are at stake). Thus, in this section, we have seen a nominal counterpart to the cases of bans on adjacent verbs in the previous section; in terms of the theory developed here, multiple DPs cannot be linearized if they are in the same strong phase.

2.2.2 Distinctness without Linear Adjacency

In the previous sections I discussed cases that showed that linear adjacency was not sufficient to trigger Distinctness effects; in cases in which a strong phase boundary intervened between the two potentially offending objects, linearization would succeed even if the objects were linearly adjacent. This was the case, for instance, in the Chaha example in (63):

(63) Gɨyə [$_{vP}$ fərəz] nəkəsəm (Chaha)
     dog   horse   bit
     'A dog bit a (nonspecific) horse'

In (63), the nonspecific direct object need not be marked with *yə-*. In the account developed above, this was because the direct object has not undergone object shift to the edge of the v_CP phase; as a result, it is linearized within the v_CP phase, and the subject is linearized in the higher CP phase. Because these two DPs are not linearized in the same Spell-Out domain, no Distinctness violation is incurred. The fact that the two DPs are linearly adjacent is irrelevant. In this section I will show that linear adjacency is not only not sufficient but also not necessary to trigger Distinctness effects; syntactic objects that are not linearly adjacent can still exhibit a Distinctness effect.

One of the other things we saw in the last section was that one way of circumventing Distinctness is to add more functional structure. In languages like Chaha, Hindi, Miskitu, and Spanish, for instance, one way of circumventing Distinctness violations is to embed one of two potentially offending DPs in a KP; I hypothesized that this KP is a phase, insulating the DP it dominates from other DPs.

Of course, the functional structure that was added in the cases discussed above sometimes also had the effect of breaking up linear adjacency between potentially offending objects, and one could imagine

theories that made crucial use of this fact. In this section I will try to show
that this would be a mistake. We will see that material that linearly intervenes between two unlinearizable objects is not sufficient in and of itself
to rescue the structure.

2.2.2.1 Adverbs   Recall from the discussion in section 2.1.1 that English
and French both have processes of invesion that exhibit a Distinctness
effect; they are blocked from occurring if two DPs would appear to the
right of the verb:

(64) a. *"It's cold," told [John] [Mary]

   b. *Je me demande oú   mange [Marie] [sa   pomme] (French)
      I me ask   where eats   Marie her apple
      'I wonder where Marie eats her apple'

These facts are unaffected by the distribution of adverbs (French data
from Michel DeGraff, personal communication):

(65) a. *"It's cold," told [John] **sadly** [Mary]

   b. *Je me demande oú   mange [Marie] **habituellement**
      I me ask   where eats   Marie usually
      [sa   pomme] (French)
      her apple
      'I wonder where Marie usually eats her apple'

This is what we expect. Consider a tree for the v_CP phase of (65a):

(66)

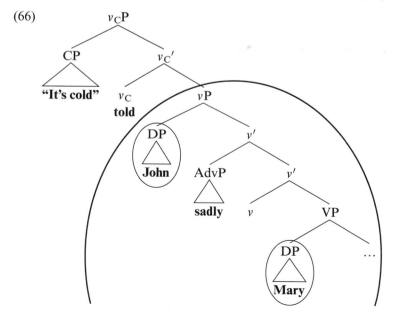

Here I have represented the quote "It's cold" as a simple CP, though we could certainly consider more complex representations. I have made the adverb *sadly* a vP adjunct, though a variety of possible placements for it are consistent with the point being made here. Even if we thought that AdvP was a phase, this AdvP (unlike the KP projections in the differential case-marking examples) does not dominate either of the offending DPs. As a result, even if Adv^0 turns out to be a phase head, this phase boundary will not protect the two DPs from being linearized in the same Spell-Out domain, and we correctly expect the examples to be ill-formed.

2.2.2.2 Relativization   Kayne (1977), Chomsky (1977, 1980), Cinque (1981), Pesetsky (1998), and Pesetsky and Torrego (2006), among many others, discuss a pattern of relativization in various Romance languages and in English infinitival relatives. In these relative clauses, a PP may appear as a relative operator, but a DP cannot:

(67) a. a person [with whom to dance]
  b. *a person [whom to admire]
  c. a person [to admire]

(68) a. l'homme [avec qui   j'ai   dansé] (French)
        the-man with whom I-have danced (Pesetsky 1998, 341)
  b. *l'homme [qui   je connais]
        the-man   whom I know
  c. l'homme [que je connais]
        the-man that I know

Chomsky's (1977, 1980) classic account of these facts was in terms of "recoverability up to deletion." On this account, a DP relative operator like the ones in (67b) and (68b) is required to delete because it can be recovered from the context, while PP operators, not being recoverable, are not required to delete.

As Pesetsky (1998) notes, data like those in (69) raise a potential problem for this kind of account:

(69) a. *a person [whose uncle to admire]
  b. l'homme [la   femme de qui   tu   as
        the-man   the wife   of whom you have
        insultée] (French)
        insulted (Pesetsky 1998, 343)

In the relative clauses in (69), the relative operator presumably contributes information that cannot be recovered if the operator is deleted. "Deletion up to recoverability," then, should allow these operators to escape deletion, which seems to be the wrong prediction.

An alternative account, which fits naturally into the theory under development here, would say that the relative operator in these examples cannot be a DP because this would bring the DP operator unacceptably close to the D of the relative clause's head, making linearization difficult. Consider the tree in (70) for the ill-formed example in (69a):

(70)   * DP

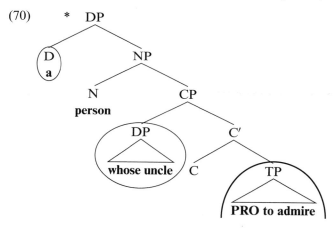

Here the relative operator *whose uncle* is in the highest specifier of the CP phase. In the version of Spell-Out that we assume here, this means that the DP *whose uncle* is not spelled out with the rest of its CP, but rather with the next higher Spell-Out domain. If we assume that DP is not a phase, then the D *a* is not a phase head, and does not trigger Spell-Out of its complement. As a result, the D *a* and the DP *whose uncle* (or, if we prefer, the D that presumably makes up part of *whose*) are linearized in the same Spell-Out domain, and the resulting linearization statement ⟨D, D⟩ is uninterpretable. Deleting the relative operator would avoid the linearization problem;[19] the next closest DP to the higher D is the PRO subject of the relative clause, but these two instances of D are separated by a Spell-Out boundary.

On the other hand, if the relative operator is a PP, the structure will be linearizable, as long as PP is a phase (as we assumed also in section 2.1.3):

(71)

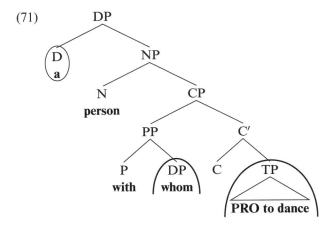

Since PP is a phase, it will spell out its complement, shielding the DP *whom* from the higher D *a*.

Besides being another case of Distinctness, these relative-clause examples demonstrate that Distinctness effects need not be triggered by string-adjacent syntactic objects. In (70), the two instances of D are not string-adjacent.

On the account given here, the behavior of English tensed relative clauses becomes problematic, since the constraints discussed above do not hold of these relative clauses:

(72) a. the man [whom I admire]
     b. the man [whose wife I insulted]

One option would be to posit structure intervening between the head and the operator that is not present in the relative clauses discussed above, so that the relative operators in these examples are not spelled out as part of the same Spell-Out domain as the D of the modified DP.[20] We might, for example, adopt the proposal of Bianchi (1999), who offers arguments that finite relative clauses have more layers in the CP field than infinitival ones do. Working in the tradition established by Rizzi (1997), she assumes that CP should be decomposed into several functional projections with different heads. Her particular structure for tensed relative clauses in English is the one in (73):

(73)

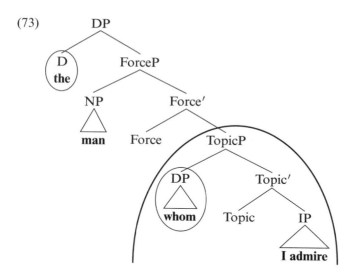

Thus, tensed relative clauses in English, for Bianchi, involve not just a CP, but a ForceP along with a TopicP, and the overt relative operator *whom* occupies the lower of these positions.

For Bianchi, the DP *whom* and the NP *man* begin the derivation as a constituent, as the object of *admire*; subsequently, this constituent moves into the specifier of TopicP, and then the NP *man* moves further into the specifier of ForceP. Regardless of whether she is right about this, the representation that she offers in (73) is one that will allow us to linearize the structure. We have already claimed that CP is a phase; once we begin considering more finely detailed structures for CP like the one in (73), we must decide which of these heads is the actual phase head. If we decide that it is the highest one, ForceP, then the structure is linearizable; the DP operator in the specifier of TopicP will be spelled out as part of the complement of ForceP, and the higher D will be spelled out in the higher Spell-Out domain.

In infinitival relative clauses, Bianchi argues that TopicP is not present. She notes, in support of this claim, that topicalization is generally unavailable in infinitival clauses in English (Bianchi 1999, 206):

(74) a. *I wonder [whether [next year PRO to go home or not]]
b. I wonder [whether [next year I'll go home or not]]

Bianchi concludes that infinitival relatives, like English infinitival clauses generally, lack a TopicP projection; the relative operator and the head noun are both, she claims, in the specifier of ForceP:

(75)

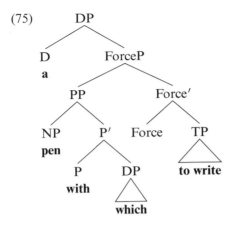

Here, for Bianchi, the PP *with which pen* has begun the derivation inside the relative clause, and has moved to the specifier of ForceP; subsequently, the NP *pen* moves into the specifier of the PP. I will have to refer interested readers to Bianchi 1999 for arguments for this conclusion. All we need to borrow from Bianchi is her argument that infinitival relatives are functionally impoverished, compared to tensed relatives in English; as a result, she argues, the relative operator in a tensed relative is more deeply buried than the corresponding operator in an infinitival relative. Bianchi's representations for relative clauses, together with the assumption that ForceP is the particular part of CP that is a phase, allow us to capture the facts.

2.2.2.3 DP-Internal Syntax Crosslinguistically, nominal arguments of nouns are often accompanied by functional structure not found with nominal arguments of verbs. In English, for instance, the complement of a noun is introduced by *of*:

(76) a. They destroyed the city
    b. *[the destruction the city]
    c. [the destruction **of** the city]

The Chaha morpheme *yə-*, which we encountered in section 2.2.1.2 as a marker of indirect objects of ditransitives and specific animate direct objects of monotransitives (a distribution for which I tried to provide an account), also appears on possessors, without regard to specificity or animacy:

(77) yə-βet   wəka (Chaha)
        yə-house roof.beam
        'the house's roof beams'

Similarly, Torrego (1998, 40) notes that complements of derived nominals
in Spanish can sometimes be marked with *a*, and that such marking does
not exhibit the animacy restriction typical of *a* on direct objects of verbs:

(78) a. Su amor al   dinero (Spanish)
          his love *a*-the money
          'his love of money'
      b. Aman (*a) el   dinero
          they.love *a* the money
          'They love money'

We will return in section 2.4.1.1 to the question of why animacy is no
longer relevant when these markers are used on DP-internal arguments.

   Some of these facts about the arguments of nouns are among those that
have classically been attributed to the Case Filter; nouns are supposed to
be unable to license Case. A Distinctness-based account might be able to
provide more insight into these facts. The account would have the virtue,
unlike previous Case-based accounts, of offering an explanation for why
it is <u>nouns</u> that require additional structure on their nominal complements. Consider a tree for the ill-formed (76b):

(79)   *   DP

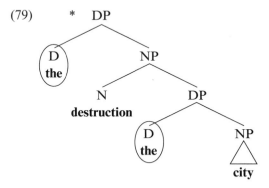

This tree is unlinearizable, for reasons that by now are familiar; in the
process of linearizing it, the grammar will generate the ordered pair
$\langle D, D \rangle$, which will cause the derivation to crash. Again, the offending
nodes are not linearly adjacent, but they are structurally close enough together to prevent linearization from succeeding.

The solution apparently involves adding a preposition, which we must regard, on this account, as a phase head (recall that we also had to posit a PP phase in sections 2.1.3 and 2.2.2.2):

(80)

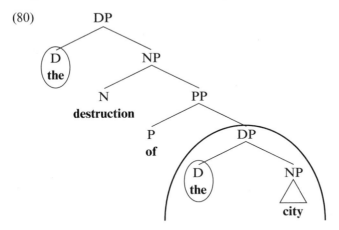

If the P *of* is a phase head, then its complement will be spelled out, and as a result, the two instances of D will be spelled out in different Spell-Out domains.

On the account developed here, the contrast in (81) and the one in (82) have the same explanation:

(81) a. *the destruction the city
     b. the destruction **of** the city

(82) a. *a man [who to dance with]
     b. a man [**with** whom to dance]

Examples (81a) and (82a), on this account, have the same problem; they both have a DP contained inside another DP, and not separated from it by any Spell-Out boundaries. (81b) and (82b) both solve this problem in the same way, by further embedding of one of the DPs. By hypothesis, the embedding of the contained DP in a PP puts it in a separate Spell-Out domain from the containing DP, rendering the structure well formed.

The contrast in (81), of course, is one of the contrasts that has classically been attributed to Case theory; nouns are unable to assign Case, for some reason, and hence cannot take DP complements. In the classic account, the contrasts in (81) and (82) have nothing to do with each other. The account developed here explains why it is nouns and not verbs that have the contrast in (81); DPs cannot be complements of nouns because this brings them unacceptably close to another instance of D. The contrast in (81) thus follows from more general considerations, which have nothing to do with Case, or even specifically with DPs.

2.3 What Nodes Are Distinct?

Distinctness outlaws linearization statements that attempt to linearize nondistinct nodes. The preceding discussion has concentrated on cases in which the relevant notion of distinctness for nodes can be captured via the rough definition offered in the introduction: two nodes are nondistinct if they have the same label. As I remarked in the introduction, this grants labels a suspicious level of importance; it is not clear that we want the grammar to refer to node labels quite this directly. I suggested there that nodes are in fact identified by their features, and that in some cases this is virtually equivalent to identifying them by their labels; in some languages, for some kinds of nodes, all nodes with the same label apparently have the same features. In this section I will go into the question of how to define distinctness a little more closely. I will consider a number of cases in which Distinctness effects are avoided by giving different features to the potentially offending nodes.

2.3.1 Polish Complementizers and Clitics

Polish has a kind of agreement morphology that can appear in a number of places (often analyzed as a clitic; for discussion see Borsley and Rivero 1994, Embick 1995, and Szczegielniak 1997, 1999). This agreement morphology appears most frequently on the complementizer or the verb (though it can also appear on focused constituents):

(83) a. On wie   że   poszedł**eš** do kina (Polish)
        he knows that went-**2SG** to movies (Adam Szczegielniak, p.c.)
        'He knows that you went to the movies'
    b. On wie   że**š**   poszedł do kina
        he knows that-**2SG** went   to movies

A colloquial register of Polish also allows declarative clauses to have two complementizers. When two complementizers appear, however, agreement must be placed on one of them (in fact, it must appear on the second complementizer, a fact Szczegielniak (1999) discusses):

(84) a. On wie   że   że**š**   poszedł do kina (Polish)
        he knows that that-**2SG** went   to movies
    b. *On wie   że   że   poszedł**eš** do kina
        he knows that that went-**2SG** to movies

This state of affairs seems amenable to an account in terms of Distinctness. In (84b), we would say, the two Cs cannot be linearized, since linearization statements like ⟨C, C⟩ make linearization crash.[21] In (84a), by contrast, one of the Cs has the agreement clitic adjoined to it. As a result, this C now has a different set of features, which means that the two C nodes are distinct. The relevant linearization statement is now something like ⟨[C], [C, 2SG]⟩. Here, then, is a particularly straightforward case of making two nodes distinct by adding features to one of them.

The Polish case has another interesting property; the requirement that agreement appear on the lower complementizer remains even if a topicalized NP linearly intervenes between the two complementizers (Szczegielniak 1999 and personal communication):

(85) a. On myšlał że   Janowi žeš   dał książkę (Polish)
         he thought that John   that-2SG gave book
         'He thought that you gave the book to John'
     b. *On myšlał że   Janowi że   dałeš   książkę
         he thought that John   that gave-2SG book

Here we have another case of a Distinctness effect that is demonstrably not about linear adjacency; the two complementizers in (85) are not linearly adjacent, but must still be distinguished by the agreement clitic. This is what we expect, since the linearly intervening material does not introduce a node that dominates the lower C, in this case, and therefore cannot possibly be introducing a phase boundary between them.[22] The relevant tree would be the one in (86):

(86)

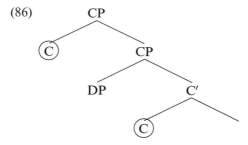

Here the linearly intervening DP is of no use in linearizing the two Cs.

2.3.2 Multiple Sluicing Revisited

We saw in section 2.1.1 that English allows multiple sluicing, as long as the *wh*-phrases obey Distinctness:

(87) a. Everyone was dancing with somebody, but I don't know [who] [with whom]

b. *Everyone insulted somebody, but I don't know [who] [whom]

Multiple sluicing in English cannot involve multiple DP remnants.

Interestingly, this condition is not universal. In German (Sauerland 1995), Japanese (Takahashi 1994), Dutch (Merchant 2001), and Greek (Merchant 2001), for example, multiple sluicing allows multiple DP remnants:

(88) a. Ich habe jedem Freund ein Buch gegeben, aber ich weiß nicht
I have every friend a book given but I know not
mehr wem welches (German)
more who which
'I gave every friend a book, but I don't remember anymore who which'

b. Watashi-wa dono otokonoko-ni-mo hoshigatteita subete-no
I-TOP every boy-DAT wanted every
hon-o ageta ga, dare-ni nani-o ka wasureta (Japanese)
book-ACC gave but who-DAT what-ACC Q forgot
'I gave every boy all the books he wanted, but I've forgotten who what'

c. Iemand heeft iets gezien, maar ik weet niet wie
someone has something seen but I know not who
wat (Dutch)
what
'Someone saw something, but I don't know who what'

d. Kapjos idhe kapjon, alla dhen ksero pjos
someone.NOM saw someone.ACC but not I.know who.NOM
pjon (Greek)
who.ACC
'Someone saw someone, but I don't know who whom'

Japanese, German, and Greek also differ from English in having comparatively rich case morphology on nominals (though this is not true of Dutch), and we might entertain the possibility that this is a relevant difference; thanks to the variety of cases in which DPs may appear in these languages, DPs are sufficiently distinct from one another that they can legitimately be brought close together.

In fact, we can find evidence that this is correct for Japanese. Consider
the contrast in (89) (Takako Iseda, Shinichiro Ishihara, Sachiko Kato,
and Kazuko Yatsushiro, personal communication):

(89) a. [Sensei-o   hihansita] gakusei-ga koko-ni oozei iru kedo,
         teacher-ACC criticized student-NOM here-DAT many be but
         dare-ga   dare-o   ka oboeteinai (Japanese)
         who-NOM who-ACC Q remember-NEG
         'There are lots of students here who criticized teachers, but I
         don't remember who who'
     b. *[Sensei-ga   suki na] gakusei-ga koko-ni oozei iru kedo,
         teacher-NOM like   student-NOM here-DAT many be but
         dare-ga   dare-ga   ka oboeteinai
         who-NOM who-NOM Q remember-NEG
         'There are lots of students here who like teachers, but I don't
         remember who who'

Although, as we have seen, Japanese does allow multiple DP remnants in
multiple sluicing, these remnants cannot all be of the same case. Example
(89a), with one nominative remnant and one accusative remnant, is well
formed, but (89b), with two nominative remnants, is ill-formed.[23] Interestingly, the facts seem to change, at least for some speakers, when the
two nominative remnants differ in animacy; these speakers find (90) better
than (89b):

(90) [Doobutsu-ga suki na] hito-ga   koko-ni oozei iru kedo,
     animal-NOM like   person-NOM here-DAT many be but
     dare-ga   nani-ga   ka oboeteinai (Japanese)
     who-NOM what-NOM Q remember-NEG
     'There are lots of people here who like animals, but I don't
     remember who what'

An anonymous reviewer points out that Japanese exhibits a similar
restriction on multiple clefting. Japanese does allow multiple clefting
(Kuwabara 1996, Koizumi 2000, Hiraiwa and Ishihara 2002, and the
references cited there):

(91) [Taroo-ga   ageta no] -wa **Hanako-ni ringo-o**   da (Japanese)
     Taroo-NOM gave C TOP Hanako-DAT apple-ACC COP
     (literally) 'It is apples to Hanako that Taroo gave'

The reviewer notes the contrast in (92), which takes advantage of the fact that the predicate *suki* 'like' may either mark both of its arguments nominative or may mark one nominative and the other accusative:

(92) a.   [Suki na no] -wa **Taroo-ga**   **Hanako-o**   da (Japanese)
        like COP C TOP Taroo-NOM Hanako-ACC COP
        (literally) 'It is Taroo Hanako that likes'
    b. ??[Suki na no] -wa **Taroo-ga**   **Hanako-ga**   da
        like COP C TOP Taroo-NOM Hanako-NOM COP
        (literally) 'It is Taroo Hanako that likes'

Thus, the reviewer points out, multiple clefts in Japanese are generally allowed, but not when the clefted phrases are DPs with the same case, as in (92b).[24]

Japanese exhibits a similar restriction on multiple rightward shift. Although Japanese is head-final, it can shift constituents to the right of the verb (see section 3.5.2 for more discussion of this phenomenon):

(93) John-ga   suki da yo, **Mary-o** (Japanese)
    John-NOM like COP ASST Mary-ACC
    'John likes her, Mary'

Multiple DPs may be shifted to postverbal positions, but these must differ either in case or in animacy:

(94) a.   Suki da yo, **John-ga**   **Mary-o** (Japanese)
        like COP ASST John-NOM Mary-ACC
        'John likes Mary'
    b. Suki da yo, **John-ga**   **chokoreeto-ga**
        like COP ASST John-NOM chocolate-NOM
        'John likes chocolate'
    c. *Suki da yo, **John-ga**   **Mary-ga**
        like COP ASST John-NOM Mary-NOM
        'John likes Mary'

Example (94c) is ruled out because the two postverbal DPs are both nominative and both animate; examples in which only one DP is nominative (94a) or animate (94b) are well formed.

Apparently, then, Japanese DPs are more distinct from each other than English DPs are. Two English DPs always trigger a Distinctness violation if they are linearized in the same Spell-Out domain, as in the multiple sluicing examples. The hypothesis pursued in this chapter has been that English DPs cannot be distinguished from each other by linearization statements. As a result, if a Spell-Out domain in English contains two DPs in an asymmetric c-command relation, the two DPs will be related by a linearization statement ⟨DP, DP⟩, and this statement will cause the derivation to crash.

In Japanese, by contrast, two DPs are undistinguishable only if they have the same values for case and animacy. Linearizing multiple Japanese DPs, then, involves linearization statements like ⟨[DP, NOM, Animate], [DP, ACC, Inanimate]⟩, and such linearization statements are not self-contradictory. Crucially, the fact is not that Japanese is simply immune to Distinctness; rather, its DPs apparently come in more varieties than their English counterparts.[25]

The German situation may be similar, though here the facts are much less clear-cut.[26] Unlike Japanese, German has no class of predicates that routinely assigns the same case to multiple arguments. We must therefore consider biclausal cases, with multiple sluicing remnants from different clauses. Not all speakers allow this, even when the lower clause is a restructuring infinitive, and for those who do, not all speakers find any contrasts of the type Distinctness leads us to expect. When there is a contrast, however, it runs in the expected direction:

(95) a.   Es ist einem Ritter gelungen, einen Riesen totzuschlagen,
      it is a.DAT knight succeeded a.ACC giant kill.INF
      aber ich weiß nicht mehr **welchem Ritter welchen**
      but I know not more which.DAT knight which.ACC
      Riesen (German)
      giant
      'A knight succeeded in killing a giant, but I don't know any more which knight which giant'

    b.   ??Es ist einem Ritter gelungen, einem König zu helfen, aber ich
      it is a.DAT knight succeeded a.DAT king to help but I
      weiß nicht mehr **welchem Ritter welchem König**
      know not more which.DAT knight which.DAT king
      'A knight succeeded in helping a king, but I don't know any more which knight which king'

    c.   Es ist einem Ritter gelungen, einer Königin zu helfen, aber
      it is a.DAT knight succeeded a.DAT king to help but
      ich weiß nicht mehr **welchem Ritter welcher Königin**
      I know not more which.DAT knight which.DAT queen
      'A knight succeeded in helping a queen, but I don't know any more which knight which queen'

There are speakers who accept all of the examples in (95), and speakers
who reject them all, but for the speakers who get a contrast, the contrast
is as represented here. As we expect, the least acceptable example is the
one with multiple DPs of the same gender and case.

In Greek, we find what appears to be an instance of Distinctness fed by
syncretism.[27] We have already seen that Greek allows multiple DP remnants in Sluicing (Merchant 2001):

(96) Kapjos   idhe kapjon,   alla dhen ksero pjos
     someone.NOM saw someone.ACC but not   I.know who.NOM
     pjon. (Greek)
     who.ACC
     'Someone saw someone, but I don't know who whom'

Neuter nominals in Greek have the same form in the nominative and accusative cases. When the subject and object of a verb are both neuter, the
resulting sluices are unacceptable. Thus, (96a), which has the neuter subject *kathe agori* 'every boy' and the neuter object *ena parathiro* 'one win-
dow', contrasts with (96b), in which the subject is the masculine *kathe
andhras* 'every man' (Sabine Iatridou, personal communication):

(97) a. *Ksero oti kathe agori   espase ena parathiro,   alla dhen
        I-know that every boy-NEUT broke one window-NEUT but not
        ksero   pio   pio (Greek)
        I-know which-NEUT which-NEUT
        'I know that every boy broke one window, but I don't know
        which which'
     b. Ksero oti kathe andhras   espase ena parathiro,   alla
        I-know that every man-MASC.NOM broke one window-NEUT but
        dhen ksero   pios   pio
        not I-know which-MASC.NOM which-NEUT
        'I know that every man broke one window, but I don't know
        which which'

The ill-formedness of (97a) is not simply a matter of the identical form of
the *wh*-phrases, since we find the same contrast in (98):

(98) a. *Ksero oti kathe agori   espase ena parathiro,   alla dhen
        I-know that every boy-NEUT broke one window-NEUT but not
        ksero   pio   agori   pio   parathiro (Greek)
        I-know which boy-NEUT which window-NEUT
        'I know that every boy broke one window, but I don't know
        which boy which window'

b. Ksero oti kathe andhras   espase ena parathiro,   alla
    I-know that every man-MASC.NOM broke one window-NEUT but
    dhen ksero pios andhras   pio   parathiro
    not I-know which man-MASC.NOM which window-NEUT
    'I know that every man broke one window, but I don't know
    which man which window'

In the introduction to this chapter, I suggested that Distinctness effects
arise because linearization applies to a representation that has not yet undergone full lexical insertion of functional heads; consequently, functional
heads are often indistinguishable from each other, which triggers the difficulties for linearization that we have been investigating. If we are to capture these Greek facts, syncretism must involve manipulation of the
syntactic representation prior to lexical insertion, so that (in this instance)
Neuter nominative and Neuter accusative are the same. For instance, we
might posit an operation of Impoverishment, like the one in Distributed
Morphology (Bonet 1991), which deletes the Case feature (or some part
of a complex Case feature set) in the presence of the feature Neuter.

On the other hand, a reviewer notes that case syncretism in German
does not trigger Distinctness violations, offering as evidence the example
in (99):

(99) Ein   Auto hat ein   Haus zerstört, aber ich weiß nicht mehr
     a.NEUT car   has a.NEUT house destroyed but I   know not   more
     welches   Auto welches   Haus (German)
     which.NEUT car   which.NEUT house
     'A car destroyed a house, but I don't know any more which car
     which house'

In German (99), as in Greek (98a), the subject and object of the sluiced
clause are both neuter, and hence have the same case morphology in
both nominative and accusative. If the contrast between (99) and (98a) is
a genuine difference between the languages, it may be evidence that German syncretism is unlike Greek syncretism in being induced after lexical
insertion (perhaps by insertion of default morphemes, rather than by an
operation of Impoverishment). I will have to leave the matter here for
the time being, hoping to return to the question in future work.

A reviewer notes that German and Greek, in addition to allowing multiple DP sluicing, are also among the languages which Alexiadou and
Anagnostopoulou (2001, 2007) describe as allowing multiple DPs inside
the vP. As the reviewer points out, this is something we might be able to

use Distinctness to capture; a language that allows multiple DP sluicing, by hypothesis, is a language that can distinguish between different kinds of DPs, and a language that can do this ought to also be able to leave multiple DPs inside the *v*P, as long as they are DPs that the language can distinguish from each other. What we expect, then, is that the availability of (for example) multiple sluicing with DP *wh*-phrases ought to pattern together with the ability to leave multiple DPs inside *v*P. We saw in sections 2.1.1 and 2.1.5 that English allows neither of these:

(100) a. *I know everyone insulted someone, but I don't know [who] [whom].
      b. *"It's cold," told [John] [Mary]

As we have just seen, German and Greek do allow multiple sluicing with DP *wh*-phrases:

(101) a. Ich habe jedem Freund ein Buch gegeben, aber ich weiß nicht
         I   have every friend a   book given   but I   know not
         mehr wem welches (German)
         more who which
         'I gave every friend a book, but I don't remember who which'
      b. Kapjos   idhe kapjon,   alla dhe ksero   pjos
         someone.NOM saw someone.ACC but not I.know who.NOM
         pjon (Greek)
         who.ACC
         'Someone saw someone, but I don't know who whom'

Moreover, Greek and German have both been independently argued to allow multiple *v*P-internal DPs, again unlike English. Fanselow (2001) and Wurmbrand (2006) have argued that multiple DPs may remain within *v*P in German. In (102), for instance, a fronted *v*P may include both the subject and the object (Wurmbrand 2006):

(102) [Ein   junger Hund einen Briefträger gebissen] hat hier schon
       a.NOM young dog a.ACC mailman   bitten   has here already
       oft (German)
       often
       'It has happened often here already that a young dog has bitten a mailman'

Alexiadou and Anagnostopoulou (2001, 2007) argue that Greek also allows multiple DPs inside *v*P. In (103), they argue that the manner adverb *prosektika* 'carefully' marks the left edge of the *v*P, which contains both the subject and the object (Alexiadou and Anagnostopoulou 2007, 38):

(103) an ehi idi   diavasi [$_{vP}$ prosektika [o   Janis   to
      if has already read   carefully   the John.NOM the
      vivlio]] (Greek)
      book.ACC
      'If John has already read the book carefully'

Thus, as the reviewer notes, the correlation between the availability of multiple sluicing with DP *wh*-phrases and that of multiple *v*P-internal DPs seems promising, and is indeed predicted by the theory.

How can we determine independently whether a language is of the English type, in which all DPs are the same, or of the Japanese/German/ Greek/Dutch type, in which DPs come in distinct varieties? I will have to leave this question to future research, unfortunately. Case morphology in German, Japanese, and Greek is certainly richer than English case morphology by any reasonable definition, but principles based on morphological richness have not had a happy history in the syntactic literature, and I am hesitant to invoke a new one here. Moreover, in Japanese, at least, the relevant feature seems to be not only case but animacy, which is no more richly represented in Japanese morphology than it is in English. More work on the conditions on multiple sluicing in different languages seems to be in order. The point of this section is simply that languages can vary in which features they take into account for purposes of linearization; the reasons for this variation will have to be discovered in future work.

2.3.3 Multiple-*wh* Fronting

Languages with multiple overt *wh*-movement pose an apparent problem for Distinctness; how are Serbian sentences like the one in (74) (from Rudin 1988) linearized, given that they appear to have two DPs in close structural proximity to each other?

(104) Ko koga vidi? (Serbian)
      who whom sees
      'Who sees whom?'

In fact, it turns out that multiple fronted *wh*-phrases in Serbian and Croatian are subject to Distinctness.[28] Bošković (2001) discusses one kind of case that arguably falls under the general rubric of Distinctness; when two *wh*-phrases are phonologically identical, only one of them may front (an observation Bošković credits to Wayles Browne):

(105) a. **Šta** uslovljava **šta?** (Serbian)
what conditions what
'What conditions what?'
b. *__Šta__ **šta** uslovljava?
what what conditions

On further investigation, it turns out that there are a number of other Distinctness effects in this general domain, though the judgments are generally more subtle than the one in (105). The situation is reminiscent of the Japanese, German, and Greek facts reviewed in the last section; Serbian and Croatian apparently distinguish between DPs of different cases and genders, but if multiple *wh*-fronting would bring DPs with the same gender and case into proximity, it is avoided.

Consider, for example, the following Serbian sentences (Sandra Stjepanović, personal communication):[29]

(106) a. Kojem   je   čovjeku kojeg   dječaka mrsko
which.DAT AUX man.DAT which.GEN boy.GEN boring
pozdraviti? (Serbian)
greet.INF
'Which man doesn't feel like greeting which boy?'
b. *Kojem   je   čovjeku kojem   dječaku mrsko pomogati?
which.DAT AUX man.DAT which.DAT boy.DAT boring help.INF
'Which man doesn't feel like helping which boy?'
c. *Kojem   je   čovjeku kojoj   ženi
which.DAT AUX man.DAT which.DAT woman.DAT
mrsko pomogati?
boring help.INF
'Which man doesn't feel like helping which woman?'

In (106), the subject receives quirky dative case from the adjective *mrsko* 'boring'. Multiple *wh*-fronting is then banned if the result would bring two dative DPs together, regardless of whether they match in gender (106b,c).

The Croatian data in (107) are similar (Martina Gračanin-Yüksek, personal communication). These examples involve a modal construction in which the subject is dative and the verb is infinitival:

(107) a. ??Kojem   je   čovjeku kojem   dječaku
           which.DAT AUX man.DAT which.DAT boy.DAT
           pomoći? (Croatian)
           help.INF
           'Which man is to help which boy?'
      b.   Kojem   je   čovjeku pomoči   kojem   dječaku?
           which.DAT AUX man.DAT help.INF which.DAT boy.DAT
           'Which man is to help which boy?'

The ill-formed (107a) contrasts with (108), in which the two nouns have
different cases:

(108) Kojem   je   čovjeku kojeg   dječaka
      which.DAT AUX man.DAT which.GEN boy.GEN
      pozdraviti? (Croatian)
      greet.INF
      'Which man is to greet which boy?'

Martina Gračanin-Yüksek (personal communication) assigns an intermediate status to (109), in which the *wh*-fronted DPs are of the same case
but different genders:

(109) ?Kojem   je   čovjeku kojoj   ženi
       which.DAT AUX man.DAT which.DAT woman.DAT
       pomoći? (Croatian)
       help.INF
       'Which man is to help which woman?'

Gračanin-Yüksek finds (109) better than (107a) but worse than (108). If
(109) and (107a) do indeed contrast, unlike the parallel Serbian examples
in (106b,c), then this represents a difference between Serbian and Croatian, or perhaps a difference between the structure in (106) and that in
(107)–(109).

   Case syncretism has an interesting effect in this regard. Consider, for
example, multiple-*wh* questions involving the predicate *sram* 'ashamed'.
This predicate takes an accusative experiencer and a genitive theme:

(110) Koju   je   ženu   koje   žene   sram?
      which.ACC AUX woman.ACC which.GEN woman.GEN ashamed
      'Which woman is ashamed of which woman?'

Animate masculine nouns have the same form for the accusative and the
genitive (I will gloss the syncretic form with "GEN" in what follows). If

we change (110) by making both arguments of the predicate masculine nouns, speakers find the result ill-formed, but the effect is apparently very subtle (Martina Gračanin-Yüksek, Damir Ćavar, Sandra Stjepanović, and Željko Bošković, personal communication):

(111) ??Kojeg   je   čovjeka   kojeg   dječaka sram?
      which.GEN AUX man.GEN which.GEN boy.GEN ashamed
      'Which man is ashamed of which boy?'

To avoid ill-formedness in (111), the second *wh*-phrase is not moved:

(112) Kojeg   je   čovjeka   sram   kojeg   dječaka?
      which.GEN AUX man.GEN ashamed which.GEN boy.GEN
      'Which man is ashamed of which boy?'

As we saw before in Greek, we will apparently need to adopt an approach to syncretism that takes it seriously enough for its effects to be visible at the point in the derivation at which Distinctness applies.

We can find similar data in Russian.[30] Here, again, the judgments seem to be quite subtle, but speakers who report contrasts report the expected ones.

(113) a. **Kakomu   zhurnalistu   kakogo   diplomata**   nuzhno
        which.DAT journalist.DAT which.ACC diplomat.ACC must
        zavtra   privetstvovat'? (Russian)
        tomorrow greet.INF
        'Which journalist needs to greet which diplomat tomorrow?'
    b. ??**Kakomu   zhurnalistu   kakomu   diplomatu**   nuzhno
        which.DAT journalist.DAT which.DAT diplomat.DAT must
        zavtra   zvonit'?
        tomorrow call.INF
        'Which journalist needs to call which diplomat tomorrow?'
    c. **Kakomu   zhurnalistu   kakoj   zhenshchine** nuzhno
        which.DAT journalist.DAT which.DAT woman.DAT must
        zavtra   zvonit'?
        tomorrow call.INF
        'Which journalist needs to call which woman tomorrow?'

As we expect, examples like (113b) with multiple fronted DPs of the same gender and case are judged worse, by some speakers, than examples like (113a) in which the fronted DPs differ in case, or examples like (113c) in which they differ in gender.

The data from Serbian, Croatian, and Russian are interesting in several respects. First, they represent another Distinctness case, found in a particularly unpromising place; these languages typically front all *wh*-phrases, and routinely front sequences of DPs. As in the Japanese and German cases in the last section, however, this indicates not that Russian, Serbian, and Croatian *wh*-fronting is immune to Distinctness, but that these languages make finer distinctions among DPs than English does.

The particular distinctions that these languages are making is also interesting. Crucially, the requirement is not simply that the *wh*-phrases contain different words; they must have different cases. Moreover, the distinction between cases is not drawn simply on the basis of morphology; Distinctness violations can involve DPs of different genders but the same case, yielding structures that are either fully degraded or at least somewhat degraded.

2.4   How to Become Distinct

In this section we will see some of the methods languages use for avoiding Distinctness violations. The discussion will touch on several of the phenomena we have already discussed, and we will also see several new examples. Methods of avoiding Distinctness violations come in four main groups. First, we will see examples in which Distinctness violations are avoided by adding extra structure; given the approach developed here, we will have to regard these extra morphemes as phase heads, introducing a Spell-Out boundary between potentially unlinearizable nodes. Second, we will see examples in which Distinctness violations are avoided by removing offending structure. Third, we will review some cases in which movement operations that would create Distinctness violations are blocked. And finally, we will see examples in which movement breaks up potential Distinctness violations, moving offending nodes further apart.

2.4.1   Adding Structure

We have already seen several cases in which structure is added to avoid a Distinctness violation. The contrast in (114) is one case in point:

(114) a. *the destruction the city
          b.   the destruction **of** the city

The following sections will discuss some similar cases.

(p. 208 х 13)

2.4.1.1 Differential Case Marking In section 2.2.1.2, we saw that in some languages, case particles are added to DPs that are too close to other DPs for linearization:

(115) Gɨyə yə-fərəz nəkʷəsənɨm (Chaha)
dog **yə**-horse bit
'A dog bit a (specific) horse'

When it is being added to direct objects or indirect objects, this case particle is only added to animate DPs, a distinction Chaha shares with Spanish, Hindi, and Miskitu. In section 2.2.2.3, we saw that the animacy restriction disappears when a DP is embedded in another DP:

(116) yə-βet wəka (Chaha)
yə-house roof.beam
'the house's roof beams'

Torrego (1998) points out similar examples in Spanish:

(117) a. Su amor al dinero (Spanish)
his love **a**-the money
'his love of money'
b. Aman (*a) el dinero
they.love *a* the money
'They love money'

The theory under development here might make this look like a natural contrast.

In the chapter so far, I have drawn trees as though the extended projection of a noun phrase consists entirely of NP and DP (and sometimes K(ase)P). Much work on DP structure, however, argues that the functional structure is more articulated than that (Bernstein 1991, Ritter 1991, Valois 1991, and Longobardi 1994, 2001, among much other work). Just to avoid committing myself to any particular structure, I will simply insert an FP between DP and NP in the trees that follow, which will represent whatever functional structure intervenes between these nodes.

Let us consider trees for (117a,b). Movement traces in (118) are suppressed (and I will ignore, for the time being, the possessor *su* 'his' in (118a); see section 2.4.2.2 for further discussion of possessors):

p. 70

(118) a.

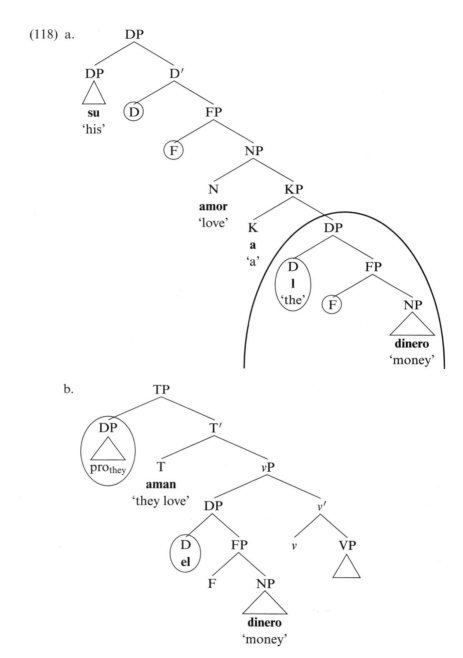

Consider the position of the DP *el dinero* 'the money' in these two trees. In (118a), this DP is asymmetrically c-commanded, not just by another

D, but also by F, which, let us recall, represents here all the functional structure in the extended projection of the embedding noun *amor* 'love'. In (118b), on the other hand, *el dinero* is asymmetrically c-commanded by another DP, but not by F or FP. The fact that *el dinero* must be embedded in a KP in (118a) but need not be in (118b) might be made to follow from this. The idea would be that Spanish DPs, like Croatian and Japanese DPs, come in more varieties than English DPs do; in particular, a DP may be associated with either animate or inanimate features. As a result, sentences like (118b) do not pose a problem for Distinctness, since the two DPs can be linearized via a linearization statement ⟨[DP, animate], [DP, inanimate]⟩.

What about (118a)? Here the DP *el dinero* 'the money' is c-commanded by F as well as by D. Thus, even if the D is safe from a Distinctness violation by virtue of the features associated with it, as long as there is some head in F which *el dinero* and the embedding DP have in common, Distinctness will be violated. If, for example, there is an *n*P parallel to *v*P, which all DPs share, then these two instances of *n* will violate Distinctness if nothing is done. Thus, although no *a* is needed in (118b), an *a* is needed in (118a). To put it another way, we expect differential Case marking to appear more consistently for DP-internal arguments than for direct objects.[31]

2.4.1.2 Gerunds, Adjective Complements

The introduction of FP in the preceding section allows us to extend the account of the behavior of English nominals to two other types of expressions: gerunds and adjectives.

A rich literature on gerunds (see Chomsky 1970; Abney 1987; Alexiadou 2001; Alexiadou, Haegeman, and Stavrou 2007, and the references cited there) has established the existence of several classes of gerunds in English, which differ from each other and from ordinary nominals in various ways. One division among gerunds has to do with the treatment of DP complements. We find gerunds in which these are marked with *of*, and gerunds in which they are not ((119a) is often called an *ing-of* gerund, while (119b) is a *PRO-ing* gerund):

(119) a. [Singing **of** the national anthem] is strictly prohibited
      b. [Singing the national anthem] is strictly prohibited

Moreover, we find instances of gerunds with overt determiners, and gerunds without them:

(120) a. [The singing] gets on my nerves
      b. [Singing] gets on my nerves

These options for gerunds do not combine freely, however. In particular, if a determiner is present, then any DP complement must be marked with *of*:

(121) a. [**The** singing **of** the national anthem] is strictly prohibited
      b. *[**The** singing the national anthem] is strictly prohibited
      c. [Singing **of** the national anthem] is strictly prohibited
      d. [Singing the national anthem] is strictly prohibited

The optionality in (121c–d) is quite odd. In general, we have seen that insertion of *of* is only permissible when it is required to avoid Distinctness; it cannot take place freely on objects of verbs, for example. Why should it be optional just here?

As is standard in the literature on gerunds, we can describe these gerunds as varying in how much nominal and verbal structure they have. In the case of (121c–d), we can say that the gerund either has, or lacks, *v*P and *v*_CP, the structures associated with ordinary transitive verbs. In particular, suppose we say that *v*P and *v*_CP are in complementary distribution with the various projections that I have been abbreviating with FP:

(122) a.

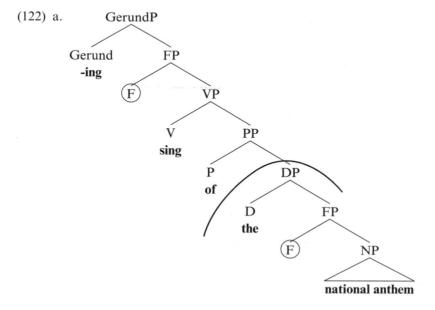

b.

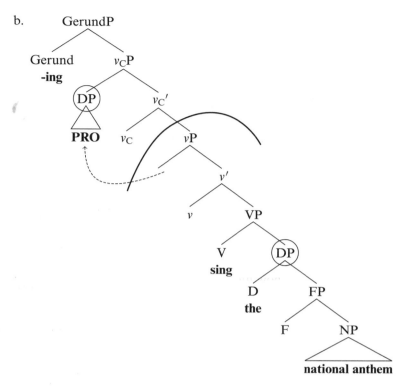

In (122b), transitive vP assigns a θ-role to its specifier, as it always does; in this case, the θ-role is assigned to a null DP PRO, which is shielded from the DP object by the Spell-Out boundary introduced by v_C. I have drawn (122a) as though there is no PRO; there could in fact be one, as long as it is high enough in the structure of the gerund to avoid being c-commanded by F.

I have drawn both of the trees in (122) as though the gerund lacks a DP layer. For (122a), it does not matter whether this is true; the tree could contain a phonologically null version of D, since the Spell-Out boundary introduced by the phase head P would protect the gerund's D from the D of *the national anthem*. The tree in (122b), by contrast, cannot have a higher D; its D would be unlinearizable with the DP PRO. We therefore expect that the tree in (122a), but not the one in (122b), may combine with an overt determiner. This is indeed what we find, as the contrast in (121a–b) demonstrates.

(123) a.

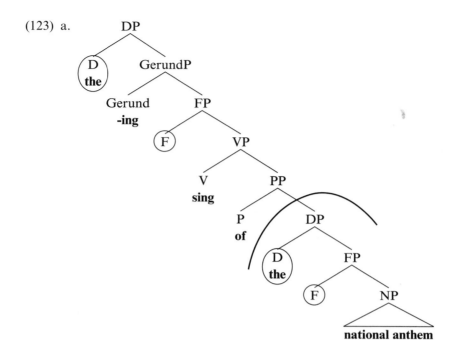

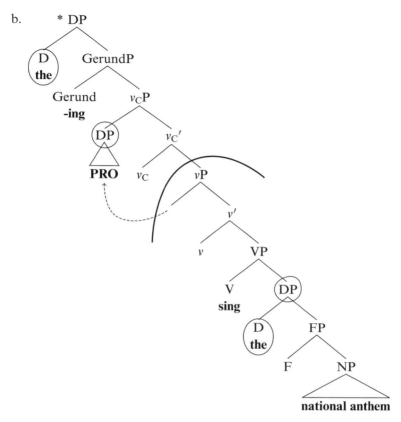

In (123a), the Spell-Out boundary introduced by the PP phase separates the functional structure of *the national anthem* from that of the gerund, and the result is well formed. In (123b), by contrast, there are three DPs in the structure and only one Spell-Out boundary; consequently, the DP PRO and the D of the gerund are in the same Spell-Out domain, and linearization fails. We thus derive the fact that if the gerund has an overt determiner, its object must be introduced with a preposition.[32]

Gerunds are also traditionally classified by how their overt agents are expressed. Agents of gerunds may be marked as possessors or may be accusative:

(124) a. We were surprised by [his singing (of) the national anthem]
      b. We were surprised by [him singing (*of) the national anthem]

As (124a) shows, when the agent of a gerund is marked as a possessor, *of* may be present or not (in traditional terms, the gerund may be either an *ing-of* gerund or a *POSS-ing* gerund). On the other hand, when the agent of the gerund is accusative, *of* must be absent (this is an *ACC-ing* gerund). In terms of the theory developed here, a genitive agent is compatible with either of the trees in (122), while an accusative one requires the tree in (122b). At least two questions then arise; why is the presence of an agent incompatible with the presence of a D head, and why is an accusative agent incompatible with either D or F?

We have seen the answer to the first of these questions in the trees in (123); if the gerund has an agent, it cannot also have a DP, since the D of the gerund and the D of the agent cannot be linearized. More generally, we can appeal to this explanation to deal with the fact that possessors and determiners are in complementary distribution in English:

(125) a. *the Mary's book
       b. *Mary's the book

Just as in (123b) above, the presence of both a determiner and a DP possessor will create insuperable problems for Distinctness. A classic account of the facts in (125) holds that *'s* is itself a determiner; we will see evidence against this account in section 2.4.2.3.

Let us now turn to the second of the questions asked above; why is it that the tree in (122a), which lacks D but has some other set of nominal projections FP, is compatible with genitive but not with accusative agents? Since this tree is compatible with a genitive agent, it must contain some position in which a θ-role can be assigned to an agent. I will create a position for this θ-role by positing a projection *n*P, parallel to *v*P (presumably one of the functional heads being abbreviated in these trees as FP):

(126)

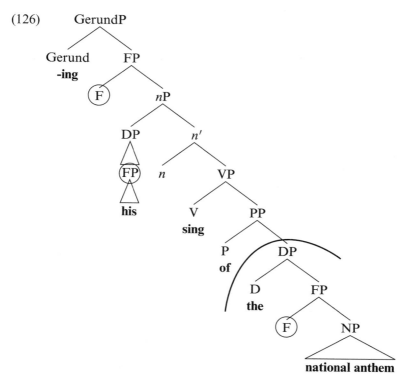

As it stands, the tree in (126) is unlinearizable; the F of the gerund c-commands the agent DP *his*, which contains an FP of its own. We could circumvent this problem by allowing the agent to move to a higher specifier, perhaps the specifier of FP:

(127)

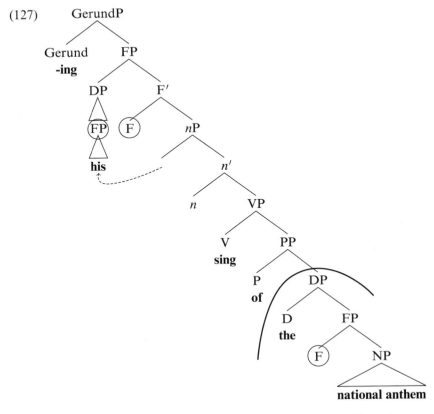

Movement of the agent in (127) repairs the linearization problem; the tree no longer contains multiple instances of F in a c-command relation. We could account for the fact that the tree in (127) is possible for genitive agents but not for accusative agents, then, by restricting the movement operation seen here to genitive agents.

As it happens, the idea that prenominal genitives move to a high position in the functional structure of the nominal has a long history in the literature on DP structure (see, for instance, Giorgi and Longobardi 1991, and the references cited there). Prenominal genitives are clearly capable of bearing a number of thematic roles, which is one argument that they are in a derived rather than a base-generated position:

(128) a. the enemy's destruction of the city
      b. the city's destruction

The idea that genitive agents of gerunds might be comparatively high in the structure thus has some independent support.

At this point it might be useful to reexamine the properties of the
"Doubl-*ing* Filter," discussed in section 2.2.1.1. We expect gerunds to exhibit the effects of this filter, and indeed they do; the examples in (129) are
based on Ross's (1972) example (20a), and the gerund in (129d) has the
tree in (129e):

(129) a. *[Keeping chanting ads] is annoying.
       b. *[The keeping chanting of ads] is annoying.
       c. *We were annoyed by [him keeping chanting ads]
       d. *We were annoyed by [his keeping chanting ads]
       e. *GerundP

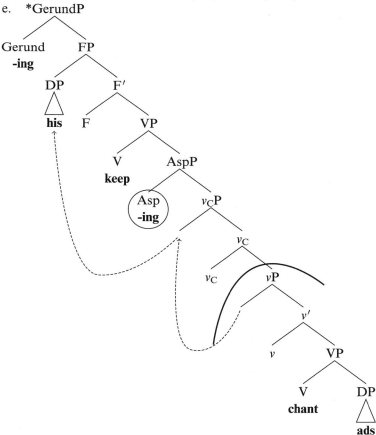

In (129e), the gerund is a raising verb, aspectual *keep*, which takes a transitive verb in a progressive form as its complement. The agent *his* begins
the derivation as the subject of *chant* and raises to the specifier of FP,
avoiding linearization together with the object DP *ads*.

On the account developed here, the problem with all of the examples in (129) is that *keep* is intransitive, taking no nominal complement. As a result, *keep* will not be separated from its complement by any Spell-Out boundaries; in particular, the *v*P associated with *keep* is not transitive, and therefore is not associated with a phase head (see section 2.2.1.1 for further discussion of this). As a result, the two instances of *-ing* are linearized in the same Spell-Out domain; I have given them different labels here, just to make the tree easy to read, but as long as linearization cannot tell them apart, the ill-formedness of these examples is expected.

By contrast, we expect that a transitive gerund with a gerundive complement will always be acceptable, even if the "Doubl-*ing* Filter" is superficially violated:

(130) a. [Stopping drinking on campus] will be tough.
      b. [The stopping of drinking on campus] will be tough.
      c. We were annoyed by [them stopping drinking on campus].
      d. We were annoyed by [their stopping (of) drinking on campus].
      e.

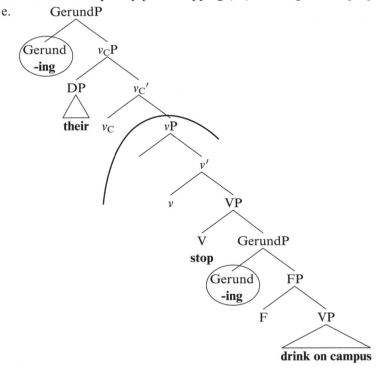

What distinguishes the examples in (130) is that the higher verb *stop* is transitive, taking a nominal complement. We have discussed various mechanisms in this chapter for protecting nominal objects of gerunds from the nominal functional structure of the gerund itself; these mechanisms will all be at work in the examples in (130). Consequently, the gerund suffix *-ing* of the complement will also be shielded from the gerund suffix *-ing* attached to *stop*, and the result will always be well formed.

Let us summarize the approach to gerunds developed here. Gerunds may in principle contain a range of the functional categories associated with nominals; we have seen evidence that they can either have or lack a DP projection, and that they may contain either the FP projections associated with ordinary nominals or the *v*P projections associated with verbs. Just when either DP or FP is present, DP complements of gerunds must be marked with *of*, to avoid Distinctness violations caused by the interaction of the functional structure of the gerund with that of its complement. If the gerund has an agent, it must lack as much functional structure as necessary to keep the agent safe from Distinctness. Genitive agents, like genitives generally, raise to a high position within FP, and therefore only require that DP be missing; accusative agents are lower in the structure, and require that both DP and FP be missing.

The introduction of FP also allows us a way of talking about the syntactic behavior of adjectives, which are like nominals in requiring insertion of *of* on their DP complements, despite not having determiners:

(131) I am fond *(of) kumquats

Although *fond* lacks a determiner, we can deal with the facts in (131) by positing some member of the FP domain in the functional structure of adjectives.

2.4.1.3 Perception Verb Passives, Italian Double Infinitives In section 2.2.1.1, we saw several cases of bans on strings of verbs, which I attributed to Distinctness:

(132) a. *John was seen __ leave
    b. *Paolo potrebbe sembrare dormire tranquillamente (Italian)
       Paolo could   seem-INF sleep-INF quietly
    c. *It's continuing raining

Several of the phenomena above have an intriguing property in common; they can be rescued by the insertion of a preposition. This is true, for instance, of some cases of passives of bare-infinitive-taking verbs in English:

(133) a. John was seen __ [to leave]
      b. John was made __ [to leave]

Here we have another instance in which a potential Distinctness violation is circumvented by adding a functional head. Given the theory developed here, we need to see these instances of *to* as phase heads (or at least as heads indicating the existence of a phase boundary).[33]

Interestingly, the insertion of *to* seems to be a last-resort strategy; the active counterparts of the examples in (133) are ill-formed:

(134) a. *We saw John to leave
      b. *We made John to leave

Similarly, in Italian, potential violations of the double-infinitive filter can be rescued by separating the two verbs with a preposition:

(135) a. *Claudio potrebbe desiderare finire   il suo lavoro (Italian)
          Claudio could   want-INF finish-INF the his work
      b. Claudio potrebbe desiderare **di** finire il suo lavoro

For some verbs, at least, this preposition insertion has the same last-resort character that it does in English; for some speakers, such prepositions cannot be naturally inserted when no violation of the double-infinitive filter is at stake:

(136) ??Claudio desidera **di** finire il suo lavoro (Italian)
       Claudio wants   to finish the his work

2.4.2 Deleting Structure

Consider the tree in (137), in which two DPs share a Spell-Out domain:

(137)

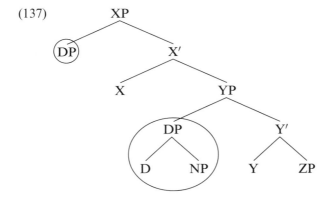

The tree in (137) is unlinearizable, since the process of linearizing it will generate the uninterpretable ordering statement ⟨DP, DP⟩. The previous sections have concentrated on one way of avoiding this kind of violation, which is to add a layer of structure to one of the DPs, protecting it from linearization with the other DP.

Another way of fixing the tree in (137), however, would be to remove functional structure from one of the DPs, making that DP into an NP:

(138)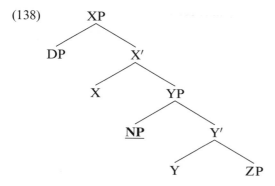

The tree in (138) is linearizable, since there is now no ordering statement ⟨DP, DP⟩.

The following sections will concentrate on cases of the type illustrated in (138), in which Distinctness violations are avoided by removing structure. Recall from the introduction to this chapter that lexical heads are very generally immune to Distinctness effects. As we will see, one response to Distinctness violations is removal of the offending functional structure.

2.4.2.1 Restructuring Longobardi (1980) offers one general class of counterexamples to the Italian double-infinitive filter, which is of interest here. Restructuring infinitives may freely violate the double-infinitive filter, and restructuring becomes obligatory when a violation of the filter is at stake:

(139) a. Giovanni comincia a voler**lo**   fare (Italian)
          Giovanni begins   to want-INF-**it** do-INF
          'Giovanni is beginning to want to do it'
       b. *Giovanni comincia a voler far**lo**

The obligatory nature of restructuring in (139) is indicated by the obligatory clitic climbing. Note that the clitic climbing is not required because the clitic must intervene between the two verbs in order for the example to be well formed; (140) is also well formed, presumably because *volere* selects for a restructuring infinitive, though there are no clitics in this example to undergo clitic climbing:

(140) Giovanni comincia a voler   viaggiare da solo (Italian)
       Giovanni begins   to want-INF travel-INF alone
       'Giovanni is beginning to want to travel alone'

Wurmbrand (1998, 2003) argues convincingly that restructuring infinitives lack functional structure, consisting simply of a VP. In her approach, a sentence like (140) would contain the partial structure in (141):

(141)

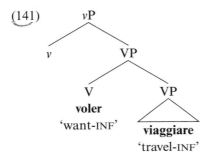

p. 33

The discussion of the double-infinitive filter in section 2.2.1.1 posited multiple *v* heads that need to be linearized. But in this case the lower verb lacks a *v* head. The pair ⟨*voler, viaggiare*⟩ can therefore be used to linearize the sentence successfully; since these are both lexical heads, by hypothesis, they undergo vocabulary insertion prior to linearization, and linearization is therefore capable of distinguishing them from each other.

2.4.2.2 Construct State   Section 2.2.2.3 discussed contrasts like the one in (142):

(142) a. *[the destruction [a city]]
       b.   [the destruction [of a city]]

The problem in (142a), on the theory developed here, was the difficulty of linearizing the D *the* with the D *a*; an ordering statement of the form ⟨D, D⟩ is ruled out by Distinctness. In (142b), on the other hand, *the* and *a* can be linearized; the PP node that dominates *of a city* introduces a phase boundary that protects the two DPs from each other, so that they are not linearized in the same Spell-Out domain and no ordering statement need order them directly:

(143) a.

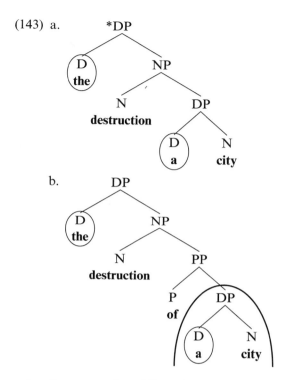

b.

Another way of avoiding the problem in (142a) would be to convert one of the DPs into an NP, removing its functional structure:

(144)

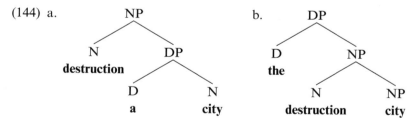

Both of the trees in (144) contain multiple instances of N, but as we have seen several times, Distinctness is entirely concerned with relations between functional heads; Spell-Out domains containing multiple lexical heads can always be linearized (for discussion, see the introduction to this chapter).

In fact, omission of the higher determiner is indeed a crosslinguistically ⌐popular⌐ approach to expressing nominal possessors, going by the name of⌐construct state. In Hebrew, for instance, possession may be indicated either with a structure like that in (142b) (given in (145a)) or with a construct state nominal like the one in (145b). The examples in (145) are synonymous (Ritter 1991):

(145) a. ha-bayit šel ha-mora (Hebrew)
          the-house of the-teacher
      b. beyt   ha-mora
          house the-teacher

A nominal like (145b) should pose no problems for linearization, since there is only one determiner.[34] Crucially, construct state nominals cannot contain two determiners:

(146) *ha-beyt ha-mora (Hebrew)

Irish also exhibits construct state nominals:[35]

(147) a.   hata an fhir (Irish)
           hat   the man.GEN
           'the man's hat'
      b. *an hata an fhir

Interestingly, in cases of multiply embedded construct state nominals, all the nouns except the most deeply embedded one not only must lack determiners but must be in the nominative (unmarked) case, rather than in the genitive case (Ken Hale, personal communication; Bammesberger 1983, 33):

(148) a.   hata fhear an tí (Irish)
           hat   man the house.GEN
           'the hat of the man of the house'
      b. *hata an fhear an ti
      c. *hata (an) fhir   an tí
           hat   the man.GEN the house.GEN

We might take this as an indication that all of the nouns other than the most deeply embedded one must lack all functional structure; not only must D be missing, but K(ase) must be missing as well. In previous examples involving K I have suggested that this is a phase head, shielding the DP it dominates from linearization with higher instances of D. Apparently in Irish this is not the case: if K were a phase head in Irish, there would be no reason to remove it in examples like (148a). It is perhaps relevant that modern Irish does not distinguish nominative and accusative case; this suggests that Case may not function as a means of linearization in Irish.[36]

Walter (2005) points out that Akkadian had a similar type of construct state, with case morphology vanishing from the head noun. Example (149a) shows the non–construct state strategy, involving insertion of a preposition, while (149b) shows construct state:

(149) a. kasp-um ša šarr-im (Akkadian)
           silver-NOM of king-GEN
       b. kasap šarr-im
           silver king-GEN
           'king's silver'

Interestingly, Walter also notes that relative clauses (which must be introduced by DP operators in Akkadian) have the same two options as possessive constructions; recall from section 2.2.2.2 that relative clauses in many languages show Distinctness effects, presumably triggered by the proximity of the relative operator to the functional structure of the head noun:

(150) a. kasp-um ša itbal-u   (šu) (Akkadian)
           silver-NOM that he.took-SUBORD it
       b. kasap itbal-u   (šu)
           silver he.took-SUBORD it
           'the silver that he took'

As with noun possession, the head noun must either be separated from the relative clause by the particle ša or must appear in a bare, caseless form. Distinctness allows the facts in (149) and those in (150) to be connected; they represent two strategies for linearizing structures with one DP embedded inside another.

We can relate these facts to another fact discussed in section 2.4.1.1; when a DP is dominated by another DP projection, all of the functional projections of the embedded DP may be c-commanded by corresponding functional heads in the higher DP. In section 2.4.1.1, this was the explanation for why differential Case marking appears on inanimate DPs just when those DPs are embedded in larger DPs:

(151) a. Su amor al   dinero (Spanish)
           his love a-the money
           'his love of money'

    b. Aman (*a) el   dinero
       they.love **a** the money
       'They love money'

In (151a), the embedded DP *el dinero* 'the money' must be distinct, not just from the D of the embedding DP, but from all the functional structure of the embedding DP (including, by hypothesis, structure that all DPs have in common, with the result that even inanimate DPs must receive differential case marking in this position). The Irish data in (148) and the Akkadian data in (149) and (150) have a similar explanation. We cannot avoid all the potential Distinctness violations by simply removing the DP of the possessor, but must remove all the potentially offending functional structure, including KP.[37]

In this section we have seen that one way to circumvent Distinctness when a nominal takes a nominal argument is to eliminate functional structure from one of the nominals. However, nothing that we have said so far guarantees that it should be the possessor that retains functional structure and the possessee that lacks it. We might expect to find examples of the opposite kind, in which the possessor is the nominal that loses its functional structure and appears as a bare N. Hungarian may offer an example of this type.

Hungarian possession may be expressed in either of two ways. In one, the possessor is marked with dative case, and appears to the left of the possessee's determiner (Szabolcsi 1994); following Szabolcsi, let us take this to mean that the possessor has moved to a high specifier within the possessed DP:

(152) Mari-nak a   kalap-ja-i (Hungarian)
     Mari-DAT the hat-POSS-PL
     'Mari's hats'

The possessor may also appear in the nominative case, which is unmarked. When it does so, it must lack a determiner (see Szabolcsi 1994 for convincing arguments for this) and appears to the right of the possessed DP's determiner (here, the optional determiner *a*):

(153) (a) Mari kalap-ja-i (Hungarian)
       the Mari hat-POSS-PL

This type of possession, then, involves stripping the possessor of its determiner and case morphology, and we might harbor the suspicion that the possessor in these examples has had all of its functional material removed.

On this view, this type of Hungarian possessive construction is like a construct state possessive construction, except that the DP with its functional material removed is the possessor rather than the possessee.

Munn (1995) discusses a construction in English that may have a similar structure. Examples like those in (154) involve what Munn calls "modificational possession":

(154) a. two men's shoes
       b. a girl's school

The string of words in (154a), for example, is ambiguous, meaning either 'the shoes of two men' (that is, four shoes) or 'two shoes of the type that men wear' (that is, two shoes). Similarly, (154b) can mean either 'the school of a particular girl' or 'a school for girls'. In other words, the determiners at the beginning of the strings in (154) may either belong to the head nouns (*shoes, school*) or to the apparent possessors of the head nouns (*men, girl*). We can further demonstrate this second possibility by considering examples like (155); here, since the determiner *every* can only combine with singular nouns, we know that it must be associated with the singular *shoe*, rather than the plural *men*:

(155) every men's shoe

In Munn's theory, an example like (155) has the tree in (156):

(156)

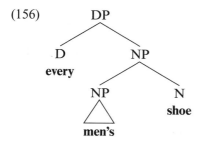

One piece of evidence for the constituent structure in (156) comes from *one*-replacement, which treats *men's shoe* as a constituent on the relevant reading:

(157) this [men's shoe] and that [one]

Modificational possession is of interest to us because it involves a possessor that crucially lacks a D, and that is left below the dominating DP rather than raising to a higher position in the noun's extended projection.

Distinctness allows us to connect these facts; if *men's* in (156) were replaced with a DP, the resulting structure would be unlinearizable, since the head of the dominating DP and that of the modificational possessor would violate Distinctness.

We can also see, by studying modificational possession, that the English genitive marker *'s* cannot always be an instance of D. In the examples of modificational possession considered above, the possessor is consistently marked with *'s*, despite the fact that the head noun has another determiner, which appears to the left of the modificational possessor. We should consider the possibility, then, that possessive *'s* is in fact never an instance of D. The classic argument that *'s* is a determiner is based on its complementary distribution with clear cases of determiners:

(158) (*the) the man's book

If we assume Distinctness, however, (158) can be ruled out without necessarily declaring *'s* to be a D. We can say that D and *'s* cannot co-occur because ordinary possessors (as opposed to modificational ones) have an instance of D of their own, and Distinctness therefore bans the co-occurrence of an ordinary possessor and an overt D for the possessed noun.

2.4.3 Movement Suppression

A third method that we find for avoiding Distinctness violations is failure to perform a movement operation that would create an unlinearizable structure. One example of this arose in section 2.3.3; we saw there that in Croatian, *wh*-movement that would bring two DPs with the same case and gender unacceptably close together is suppressed:

(159) a. ??Kojem   je   čovjeku kojem   dječaku
          which.DAT AUX man.DAT which.DAT boy.DAT
          pomoči? (Croatian)
          help.INF
          'Which man is to help which boy?'
     b.   Kojem   je   čovjeku pomoči kojem   dječaku?
          which.DAT AUX man.DAT help.INF which.DAT boy.DAT
          'Which man is to help which boy?'

Salanova (2004) discusses another case of a similar kind. The case involves inversion of the subject and the verbal complex (that is, the verb along with any auxiliaries, clitics, and negation) in Río de la Plata Spanish. In general, the subject and the verb may freely invert in Spanish:

(160) a. Juan recibió la encomienda (Spanish)
Juan received the parcel
'Juan received the parcel'
b. Recibió Juan la encomienda
c. Recibió la encomienda Juan

Wh-questions do not generally require subject-verb inversion:

(161) A quién María le mandó el paquete? (Spanish)
to whom Maria to.him sent the package
'Who did Maria send the package to?'

However, inversion becomes obligatory in two types of cases. It is obligatory when a bare DP is *wh*-moved past the subject:

(162) a. *Quién Juan quiere que le escriba? (Spanish)
who Juan wants that to.him writes
'Who does Juan want writing him?'
b. Quién quiere Juan que le escriba?

Inversion is also obligatory when the subject and the fronted *wh*-phrase are both dative:

(163) a. *A quién a Juan le pareció que le habían dado
to whom to Juan to.him seemed that to.him they.had given
el premio?
the prize
'To whom did it seem to Juan that they had given the prize?'
b. A quién le pareció a Juan que le habían dado el premio?

Inversion is not obligatory if, for example, the subject is nominative and the *wh*-fronted phrase is marked with *a*:

(164) A quién Juan conoció en Buenos Aires? (Spanish)
to whom Juan met in Buenos Aires
'Who did Juan meet in Buenos Aires?'

Salanova (2004) develops an account of these facts in terms of Distinctness. In this case, the phrases interacting for Distinctness are the fronted *wh*-phrase and the subject. Example (162a) has the tree in (165):

(165)

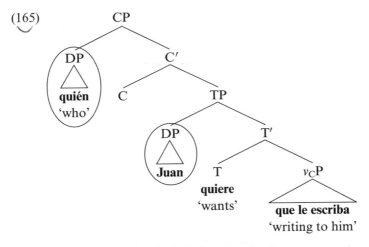

In (165) the two DPs *quién* and *Juan* are in close structural proximity. If they are linearized in the same Spell-Out domain (an issue to which I will return in a moment), then linearization ought to fail.

Salanova offers arguments that inversion in RP Spanish involves leaving the subject in a structurally lower position (rather than, for example, head movement of the verb into a higher position). The well-formed (162b), for instance, has the tree in (166):

(166)

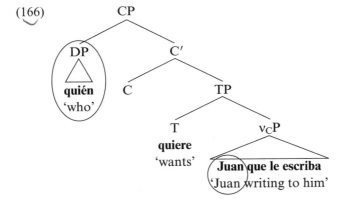

The only difference between (165) and (166) is that in (166), the subject *Juan* has failed to raise to the external subject position. If *Juan* can be spelled out in the v_CP phase, the two DPs will be linearized in different Spell-Out domains, and linearization will succeed.

As Salanova points out, this account of the ill-formedness of (165) crucially requires the specifier of CP and the specifier of TP to be spelled out in the same Spell-Out domain. He notes that the specifier of matrix CP has a peculiar status in phase theory; since it is the 'edge' of the CP phase, it ought not to be spelled out until a higher phase is completed, but in fact there is no higher phase. Salanova suggests that in RP Spanish, at least, the highest CP edge is spelled out with the rest of its phase, unlike other phase edges.[38] This assumption allows us to account for a contrast between matrix and embedded questions. Inversion is never obligatory in embedded questions in RP Spanish:

(167) Preguntó qué   Juan quería hacer (Spanish)
      he.asked what Juan wanted to.do

The lack of inversion in (167) is expected on Salanova's theory. Because the embedded CP is not the highest phase in the sentence, it can undergo Spell-Out in the ordinary way, spelling out its specifier with the higher Spell-Out domain. As a result, *qué* and *Juan* are not spelled out in the same Spell-Out domain in (167), although they would be if the embedded question were a matrix question. Thus, inversion is not obligatory in this case.

We now arrive at an interesting difficulty for our theory, however. Matrix questions with *qué* 'what' require inversion:

(168) a. Qué vio Juan en Buenos Aires? (Spanish)
          what saw Juan in Buenos Aires
          'What did Juan see in Buenos Aires?'
      b. *Qué Juan vio en Buenos Aires?

Salanova's account therefore requires *qué* 'what' and *Juan* to be treated as identical for Distinctness. In section 2.2.1.2, however, I outlined an account of differential case marking, in which I noted that this marking appears specifically on animate objects:

(169) a. Laura escondió a un prisionero durante dos años (Spanish)
          Laura hid   **a** a prisoner for   two years
          'Laura hid a (specific) prisoner for two years'
      b. Golpeó (* a) la mesa
          he/she.hit **a** the table
          'He/she hit the table'

Data like those in (169) might suggest that that we should allow Spanish to distinguish between DPs on the basis of animacy (just as, in sections 2.3.2 and 2.3.3, we saw evidence that languages like Japanese and

Croatian can distinguish between DPs on the basis of properties like Case, and possibly also animacy). In examples like (169b), on this view, the two DPs in the sentence may be linearized via a linearization statement ⟨[DP, animate], [DP, inanimate]⟩, and linearization therefore succeeds without the need for differential case marking on the object. In (169a), by contrast, both DPs are animate, and differential case marking is therefore necessary. Having made this move, however, we are left with a puzzle; why can we not linearize an example like (168b) via a linearization statement ⟨[DP, inanimate], [DP, animate]⟩? The discussion of Distinctness thus far has been entirely built on a symmetric relation; nodes are either the same or different, and they must be different if they are to be directly linearized. These Spanish facts suggest that it must sometimes be possible for the relevant relationship between nodes to be asymmetric; these seem to be cases in which α and β can be distinguished only if α precedes β.

To investigate this problem further, let us consider a wider range of facts about differential object marking in Spanish. We saw in (169) above that when the subject of a Spanish sentence is animate, a specific object is marked with *a* if it is also animate, and not if it is inanimate. The suggestion entertained above was that this is because the animate object of a sentence like (169a) is indistinguishable from its animate subject. As it stands, the account predicts that if the subject is inanimate, the animate object will again be distinguishable from it, and differential case marking will again become unnecessary. But this is false; differential case marking is only sensitive to the animacy of the object, not of the subject (Torrego 1998, 30):

(170) a. El soldado emborrachó *(**a**) varios colegas. (Spanish)
         the soldier made.drunk **a** several friends
         'The soldier got several friends drunk'
      b. El vino emborrachó *(**a**) varios invitados.
         the wine made.drunk **a** several guests
         'The wine made several guests drunk'

Moreover, if both the subject and the object are inanimate, differential case marking is impossible (Esther Torrego, personal communication):

(171) El coche aplastó (* **a**) una lata
      the car crushed **a** a   can
      'The car crushed a can'

The data in (169)–(171) are summarized in the table below:

(172)

subject	object	a appears?	examples
animate	animate	yes	(169a), (170a)
animate	inanimate	no	(169b)
inanimate	animate	yes	(170b)
inanimate	inanimate	no	(171)

The table in (172) demonstrates that the object is marked with *a* just if it is animate and specific, regardless of the animacy of the subject. The table also shows that the most straightforward Distinctness-based account of the distribution of *a* will fail; *a* does not appear simply when the subject and object have identical values for animacy.

One way of resolving this problem would be to effectively declare subjects to be honorary "animates," regardless of their actual status for animacy. Harbour (2007) offers a theory that we can use to formalize a version of this idea.

Harbour follows much work in morphology in assuming that DPs may be underspecified for φ-features. A first-person pronoun, for example, must bear the features [+author, +participant]. On the other hand, an inanimate DP, in principle, might not bear any version of the features [±author, ±participant]; such features may be left out of the representation, perhaps because their values can be deduced from the fact that the DP is inanimate. Building on the theory of Adger and Harbour (2007), Harbour argues that arguments merged in the specifier of *v*P must in fact be specified for the features [±author, ±participant]. Thus, an inanimate subject (of a transitive verb) must bear the features [−author, −participant], even though an inanimate object can be underspecified for these features.

As Adger and Harbour (2007) note, any [+participant] nominal will be animate, though an animate nominal need not bear a value for [participant]. Adapting Harbour's (2007) account of this kind of fact to our current purposes,[39] we might view this asymmetric relation between the [participant] feature and the [animate] feature in terms of a feature hierarchy, represented as a hierarchical syntactic structure:

(173) a.

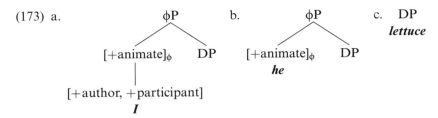

On the view represented by the trees in (173), the features [±participant] and [±author] are dependent on the feature [±animate], which is represented here as part of the head φ of a projection φP. Languages can then vary in how much of the structure not required by universal principles they project; the trees in (173) represent the smallest possible structures, which are the ones that will be useful for the analysis of Spanish. An inanimate nominal like *lettuce* might lack [animate], [participant], and [author] features, and hence have no φP at all, being dominated instead by some other set of functional heads, represented here as DP. A pronoun like *I* would be positively specified for all of the features, and would therefore have to project a φP; a third-person pronoun like *he* would be specified [+animate], and hence project a φP, but could lack a specification for [±participant, ±author].

If Harbour (2007) is correct, however, the trees in (173) cannot all appear in the specifier of *v*P, which may only be occupied by phrases specified for [±participant, ±author]. When appearing in the specifier of *v*P, then, the nominals in (173) would have the following representations:

(174) a.

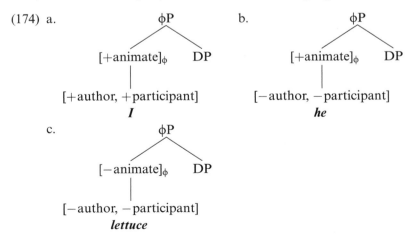

Harbour's goals have nothing to do with the Spanish facts under consideration here; he is concerned with deriving the hierarchies for case marking discussed by Silverstein (1976) and in much subsequent work.[40] Still, his theory will be useful in capturing the Spanish data. As the trees in (173) and (174) show, Harbour's theory allows us to categorize nominals into two types, depending on whether they project an φP. Animate nominals will always be specified [+animate], and will therefore always project φP. Inanimate nominals will in general have no specification for [animate] at all, and will therefore lack φP. However, when inanimate nominals appear in the specifier of *v*P, they will be required in Harbour's (2007) theory to bear the feature [−animate], and hence to project φP.

We can now rewrite the table in (172) in terms of this categorization of nominals:

(175) *subject*	*object*	*a appears?*	*examples*
φP[+animate]	φP[+animate]	yes	(169a), (170a)
φP[+animate]	DP	no	(169b)
φP[−animate]	φP[+animate]	yes	(170b)
φP[−animate]	DP	no	(99)

The table in (175) allows us to capture the distribution of *a* in terms of Distinctness. The object must be marked with *a* just when the subject and the object both bear values for [animate]; on Harbour's view, this is just when the subject and the object are both instances of φP. Harbour's theory gives us independent evidence for the idea that the distinction between animates and inanimates should be blurred just in the case of subjects; direct objects may fail to project φP, but subjects cannot, on his approach.

Now we can return to the problem in (168), repeated here as (176):

(176) a. Qué vio Juan en Buenos Aires? (Spanish)
        what saw Juan in Buenos Aires
        'What did Juan see in Buenos Aires?'
    b. *Qué Juan vio en Buenos Aires?

Wh-fronting of an inanimate *wh*-phrase triggers a Distinctness effect, realized here (following Salanova) as inversion of the subject and the verb. We have seen that when an inanimate object is not *wh*-extracted, its interaction with the subject cannot yield Distinctess effects (that is, the inanimate object cannot be marked with *a*). Why should *wh*-extraction of the inanimate object trigger a Distinctness effect?

If we take Harbour's representations for animate and inanimate nominals literally, the facts can be captured via c-command. An example like
(176a) will have a structure like the one in (177):

(177)

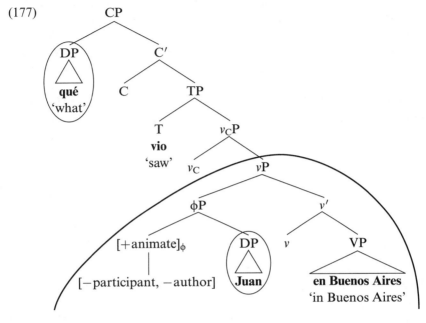

As long as Harbour's φP is the highest functional projection in the projection of Spanish nominals, and as long as it is not a phase, we expect the
subject in (177) to be required to remain in the lower phase. The *wh*-
phrase is an inanimate object, and therefore fails to project a φP, on Harbour's theory, but it does project some smaller subset of the functional
structure found in the subject, represented here as a DP. If the subject
were to raise to the specifier of TP, thereby appearing in the object's
phase, the object DP would c-command the DP inside the subject, and
Distinctness would be violated.

If the subject c-commands the object, by contrast, no c-command relation between identical functional heads is established, and Distinctness is
respected:

(178) a. Juan vio (*a) una mesa.
          Juan saw **a** a   table
          'Juan saw a table'

b.

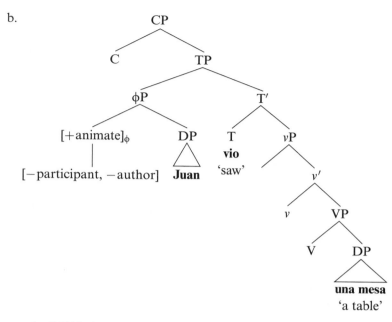

In the tree in (178b), because the subject projects φP and the object does not, no Distinctness violation arises, even if the object and the subject are in the same phase. In particular, the two instances of DP in this tree are not in a c-command relation, and hence will not create a linearization statement ⟨DP, DP⟩.

We have seen that Harbour's theory accords φP status to animate nominals, and to subjects of transitive clauses; inanimate nonsubjects, by contrast, have a more impoverished functional structure, represented here as DP. The theory developed here then leads us to expect Distinctness effects when a subject φP c-commands another φP (that is, an animate nominal); this accounts for the distribution of marking by *a*. On the hypothesis that φP is not a phase, we also expect Distinctness effects when any nominal, animate or inanimate, c-commands a subject; thus, *wh*-extraction of inanimate nominals triggers subject-verb inversion.

More generally, this account of the Spanish facts offers the hope that some of the crosslinguistic differences in the behavior of Distinctness can be linked to differences in the representation of nominal structure. In Harbour's theory, the split that we have made use of in Spanish between DP and φP represents the minimal nominal structure that must be present to satisfy universal conditions on how morphological features may be realized, but it is also crucial to his theory that languages may choose to include more structure than the universal conditions demand. On the theory developed here, we hope to discover that such variation in nominal structure will correlate with crosslinguistic differences in the behavior of Distinctness effects (for instance, the relevance of animacy in some but not all languages)

Let us close this section with another puzzle. We have seen that *wh*-fronting of an *a*-marked object does not trigger subject-verb inversion in Río de la Plata Spanish:

(179) A quién Juan conoció en Buenos Aires? (Spanish)
        to whom Juan met   in Buenos Aires
        'Who did Juan meet in Buenos Aires?'

Here the fronted object is a KP, and the subject is (on the theory presented in this section) a ϕP. Since the subject does not contain a KP, we do not expect to find Distinctness effects here, and in fact we find none; the subject and the verb need not invert.

Example (179) represents one of the ways of avoiding a Distinctness violation; the object has been converted into a KP, and is therefore linearizable with the subject ϕP. We have already seen, in Salanova's work, that Spanish has another way to avoid Distinctness violations created by *wh*-movement, which is to leave the subject inside the *v*P. Thus, we might expect to discover another way of asking the question in (179), in which the object is not marked with *a*, and the subject and the verb invert:

(180) *Quién conoció Juan en Buenos Aires?
         who   met   Juan in Buenos Aires
         'Who did Juan meet in Buenos Aires?'

Here the subject and the object are both phrases of the same type (by hypothesis, instances of ϕP), but the subject has been left in the *v*P, while the object is in a higher Spell-Out domain. Nevertheless, the result is ill-formed.

One way of characterizing the facts in (179) and (180) would be to say that Distinctness violations must be avoided whenever possible, not merely in the final PF representation but throughout the derivation. When the subject and object are both animate, a Distinctness violation will arise as soon as both subject and object are Merged into the tree:

(181)

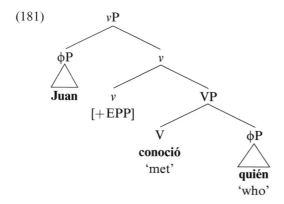

At this point in the derivation, vP contains two instances of φP, in violation of Distinctness. This violation is apparently intolerable, and must be avoided by Merging the object as a KP rather than as a φP, if this is possible. Of course, as we saw in section 2.2.1.2, the object can only be Merged as a KP if it is specific, and hence semantically capable of undergoing Object Shift; we might think of this as a requirement that the relevant K enter into an Agree relation with a higher Probe (perhaps v_C) that triggers shift of the object into a higher phase. If the object is nonspecific, the Distinctness violation in (181) must simply be tolerated until the subject can exit the phase.

If the semantic properties of the object allow it to enter the derivation as a KP, however, it must do so. The derivation cannot incur a temporary Distinctness violation by gratuitously Merging the object without K and then "waiting" for *wh*-movement to remove the object from the phase containing the subject.

If this is the correct way of dealing with these facts, it has interesting consequences for our understanding of Distinctness; rather than simply being a filter on PF representations, Distinctness must apparently be something the grammar is "aware" of, so that the derivation actively seeks to avoid Distinctness violations as it proceeds. I return to this possibility in section 4.4.2.3.5.

2.4.4 Movement

Moro (2000) develops a theory in which movement, at least in some cases, is caused by a need to avoid linearization failures. Distinctness could be used in a similar way. Since Distinctness is a condition on outputs, I will remain agnostic here about whether Distinctness simply acts as a filter, ruling out the outputs of derivations in which certain movements do not occur, or actually motivates movement in the narrow syntax.

2.4.4.1 Chol Coon (forthcoming) discusses data from Chol, a Mayan language, which are amenable to treatment in terms of Distinctness.[41] Chol sentences may have SVO word order, in which case both arguments may be DPs:

(182) **Ili** wiñik tyi i-   choñ-o   **jiñi** wakax (Chol)
      this man PERF 3ERG sell-TRANS D cow
      'This man sold the cow'

Both arguments may also be postverbal, but in this case the object must lack a determiner, and the word order must be VOS:

(183) Tyi i-   mek'-e   x'ixik   **jiñi** wiñik (Chol)
      PERF 3ERG hug-TRANS woman D   man
      'The man hugged a/the woman'

Speakers generally reject sentences with multiple postverbal DPs:[42]

(184) *Tyi i-   jats'-ä   **jiñi** ts'i' **jiñi** wiñik (Chol)
      PERF 3ERG hit-TRANS D   dog D   man
      'The man hit the dog'

The option of having a bare NP argument in Chol is reminiscent of much work on incorporation (Baker 1988 and much other work) and pseudo-incorporation (Massam 2001).[43] What is particularly interesting about Chol is the way this option is tied to word order. From a point of view that includes Distinctness, we can see the Chol facts as mirroring the facts the inversion constructions in French and English that we discussed in section 2.1.5:

(185) a. "It's cold," said John to Mary
      b. *"It's cold," told John Mary

The difference between Chol and English, on this account, has to do with the availability of a strategy in Chol that allows it to circumvent the conditions that rule out examples like (174b) in English. Coon (forthcoming) offers arguments that examples like (183) involve fronting of a phrase that contains the verb and the object (she subdivides vP into vP and VoiceP, placing the subject in the specifier of the latter):

(186)

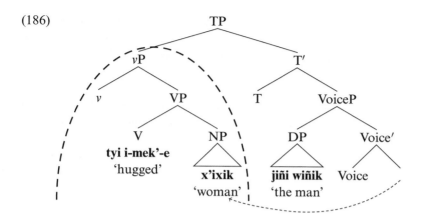

DP objects, by contrast, undergo object shift to a functional projection that Coon labels AbsP:

(187)

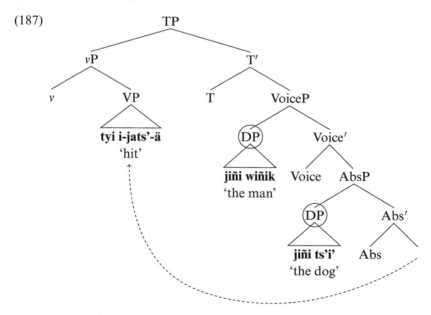

The tree in (187) has the same problems as English examples like (185b); two DPs are in close structural proximity, and linearization should fail. In (186), by contrast, movement of the *v*P has guaranteed that the two nominal arguments are not in a c-command relation with each other (moreover, even if they were, one has been stripped of its determiner). And as we saw in (182), examples in which both the subject and the object are DPs are perfectly acceptable as long as the subject has been fronted to a preverbal position (by hypothesis, a position outside the phase containing the object). We can analyze these Chol word-order patterns as being driven by the same considerations that prevent quotative inversion in (185).

2.4.4.2 Double Object Constructions   In many languages, when multiple DPs begin the derivation inside the verb phrase, one of them must be moved to a higher position. We can contrast Pylkkänen's (2002) low applicatives (exemplified in (188a)) with what I will call "prepositional datives," in which the indirect object is expressed as a prepositional phrase:

(188) a. I gave John a book
       b. I gave a book to John

Examples like (188a) have already been discussed to some extent in section 2.2.1.2, where we were trying to account for the fact that in languages like Chaha and Spanish that have differential object marking, such examples require differential object marking on the indirect object, and the absence of such marking on the direct object. In what follows we will see evidence that in some languages, one of the objects in a sentence like (188a) is required to move to a higher position, possibly crossing a Spell-Out boundary.

2.4.4.2.1 Chinese   Soh (1998), for example, discusses differences between low applicatives and prepositional-dative constructions in Chinese. The two constructions are exemplified in (189):

(189) a. wo song-le Zhangsan nei-ben   shu (Chinese)
          I   gave   Zhangsan that-CLASS book
          'I gave Zhangsan that book'
      b. wo song-le nei-ben   shu gei Zhangsan
          I   gave   that-CLASS book to Zhangsan
          'I gave that book to Zhangsan'

As Soh points out, adverbs like *liang ci* 'twice' behave differently in these two constructions. Such adverbs intervene between the two objects in the low applicative construction, but precede both of the internal arguments of the prepositional-dative construction:[44]

(190) a. wo song-le *nei-ge   pengyou* **liang ci** *xiaoshuo* (Chinese)
          I   gave   that-CLASS friend   twice   novel
          'I have given that friend a novel twice'

b. wo song-guo **liang ci** _xiaoshuo_ _gei Zhangsan_
   I   gave   twice   novel   to Zhangsan
   'I gave a novel to Zhangsan twice'

We can take these facts as evidence that the indirect object in (190a) is structurally higher than the direct object in either of the examples in (190).

2.4.4.2.2 English Emonds (1976) and Koizumi (1993) discuss data on the placement of English particles that are reminiscent of the Chinese data above. In some dialects of English, at least, postverbal particles must intervene between the two objects of a low applicative construction, but may precede both internal arguments in the prepositional-dative construction:[45]

(191) a. *The secretary sent **out** [the stockholders] [a schedule]
      b. The secretary sent [the stockholders] **out** [a schedule]

(192) a. I sent **out** [a schedule] [to the stockholders]
      b. I sent [a schedule] **out** [to the stockholders]

Both of these sets of facts seem to point toward structures in which the two internal arguments are further apart in low applicatives than they are in the prepositional-dative construction. The applied object is apparently so structurally high in the low applicative that it must precede Chinese _liang ci_ 'twice' and English postverbal particles; this seems not to be true in the prepositional-dative construction, where the DP argument may remain in a structurally low position, following the adverbial elements.

2.4.4.2.3 Kinande Baker and Collins (2006) discuss the behavior of multiple VP-internal DPs in Kinande, Ju|'hoansi, and ǂHoan. In what follows I will concentrate on the Kinande facts. The discussion will rely heavily on the facts that Baker and Collins report; Kinande data are from their paper unless labeled otherwise. I will also introduce some facts that Pierre Mujomba (personal communication) was kind enough to give me.[46] As we will see, there are some places where Mujomba disagrees with Philip Ngessimo Mutaka, the Kinande speaker who worked with Baker and Collins; given the large size of the Kinande language (roughly 4 million speakers: Mutaka and Kavutirwaki to appear), this kind of speaker variation is no surprise.

Kinande has a morpheme that Baker and Collins refer to as the "linker" (which I will gloss LI), which must appear when the VP contains multiple DPs:

(193) a. Kambale a-seng-er-a   ehilanga **hy'**
Kambale 1s/T-pack-APPL-FV 19.peanuts **19.LI**
omwami (Kinande)
1.chief
'Kambale packed peanuts for the chief'
   b. Kambale a-seng-er-a   omwami **y'** ehilanga
Kambale 1s/T-pack-APPL-FV 1.chief   **1.LI** 19.peanuts
'Kambale packed peanuts for the chief'

The linker agrees with the preceding DP in noun class. Although the two objects after the verb may be in either order, the linker must appear between them:

(194) *Kambale a-seng-er-a   ehilanga   omwami **yo** (Kinande)
Kambale 1s/T-pack-APPL-FV 19.peanuts 1.chief   **1.LI**
'Kambale packed peanuts for the chief'

Baker and Collins explain these data by depicting the linker as a functional head that attracts one of the VP-internal DPs to its specifier and agrees with it in noun class. When the VP contains two DPs, then, one must move to a higher position. Linkers do not appear when the *v*P contains only one DP:

(195) a. Kambale a-hek-er-a   omwami **y'** obwabu (Kinande)
Kambale 1s/T-carry-APPL-FV 1.chief   **1.LI** 14.drink
'Kambale carried drink for the chief'
   b. Kambale a-hek-a   (***y'***) obwabu (***bo***)
Kambale 1s/T-carry-FV **1.LI** 14.drink **14.LI**
'Kambale carried drink'
   c. Omukali mo-a-h-er-u-e   eritunda (***ryo***) na
1.woman AFF-1s/T-give-APPL-EXT-PASS 5.fruit   **5.LI** by
Kambale
Kambale
'The woman was given fruit by Kambale'

In (195b), only one DP appears after the verb, and no linker can be used; in (195c), the other *v*P-internal phrase is a PP, and again the linker cannot appear. If we are willing to regard the linker as a phase head, we can view the movement operation in (194) as Distinctness-driven; it allows us to keep the two DPs in separate Spell-Out domains. Like several of the Distinctness-avoiding mechanisms we have seen in this chapter, the linker appears to be a last-resort mechanism, appearing only when needed to fix a structure.

Much work on Kinande reaches the conclusion that Kinande preverbal subjects undergo dislocation to a structurally high position (Schneider-Zioga 2000, 2007; Baker 2003; Miyagawa, forthcoming). Schneider-Zioga (2000) notes, for example, that Kinande subjects may precede the complementizer:

(196) Kambale ng' a-langir-a Mary...
      Kambale if 1s/T-see-FV Mary
      'If Kambale saw Mary...'

Preverbal subjects in Kinande also have semantic properties associated with dislocation; for instance, a quantifier in preverbal subject position must take higher scope than a quantifier in object position (Baker 2003):

(197) Omukali a-gul-a   obuli ritunda
      1-woman 1s/T-buy-FV every 5-fruit
      'A woman bought every fruit' ∃ > ∀, *∀ > ∃

Following previous work on Kinande, then, I will place the preverbal subject in the specifier of a projection in the C domain, which I will simply label C (corresponding to the αP of Miyagawa, forthcoming).[47]

The Kinande subject need not be preverbal, however. Like a number of other Bantu languages (see, in particular, Ndayiragije 1999 on Kirundi), Kinande has a construction in which the object precedes the verb and the subject follows it, which has the effect of focusing the subject (Pierre Mujomba, personal communication):

(198) a. Omúlumy' a-ámá-hek'   akatébé (Kinande)
         1-man   1s-PRES-carry 12-bucket
         'The man carries the bucket'
      b. Akatébé   ka-ámá-hek-á   múlúme
         12-bucket 12s-PRES-carry-FV 1-man
         'The MAN carries the bucket'

I will follow the literature in calling this construction Subject-Object Reversal. If a linker appears in the postverbal field in this type of example, the subject follows the linker (Pierre Mujomba, personal communication):[48]

(199) a. Esyóngwé si-ká-seny-er'   omó músitú mó
         9-wood   9s-HAB-chop-APPL 18-in 3-forest 18.LI
         bákali (Kinande)
         2-women
         'WOMEN chop wood in the forest'

    b. *Esyóngwé si-ká-seny-er'   bákalí   b'   omó músitú
       9-wood   9s-HAB-chop-APPL 2-women 18.LI 18-in 3-forest
       'WOMEN chop wood in the forest'

I take this as evidence that the subject is generated below the linker; I
return to the question of why (199b) should be ill-formed below. Let us
assume that the linker is in fact the head v_C, the phase head that was posited in section 2.1.5 as the highest head in the vP system.

    The claims of this section are represented by the tree in (200b) for ex-
ample (193a), repeated here as (200a):

(200) a. Kambale a-seng-er-a   ehilanga   hy'
        Kambale 1s/T-pack-APPL-FV 19.peanuts 19.LI
        omwami (Kinande)
        1.chief
        'Kambale packed peanuts for the chief'

    b.

The derivation of (200b) begins with a *v*P that violates Distinctness, containing three DPs. Various aspects of the derivation are not indicated here. For instance, the subject may very well stop in the specifier of TP on its way up the tree; this will not be important in what follows. Also, I have not tried to represent head movement; in fact, if the linker is itself a functional head, the movement of the verb past the linker represents a problem for the Head Movement Constraint. Perhaps more importantly, I have made no claims about how conditions on locality of movement are satisfied in this derivation (a topic I return to later). For our purposes at the moment, the important property of the tree in (200b) is simply that by the end of the derivation, Distinctness is respected; two of the three DPs have moved to specifiers of phase heads, and each DP is in a separate Spell-Out domain.

Two types of phrases in Kinande complicate this picture in interesting ways. The first are locative expressions, and the second are a class of nominals with nonspecific indefinite interpretations.

Locatives in Kinande involve nominals preceded by morphemes like *oko* and *omo*, which have their own specifications for noun class, independent of the noun to which they are attached (*oko* is class 17, and *omo* is class 18). As we will see, nominals preceded by *omo* can mean a variety of things, including locations and instruments, but I will follow Baker and Collins in calling them "locative expressions" throughout, regardless of their semantics. Baker and Collins argue that locative expressions in Kinande have many but not all of the properties of DPs. They can, for example, trigger subject agreement on the verb in locative inversion:

(201) Oko-mesa   kw-a-hir-aw-a   ehilanga (Kinande)
     17.LOC-table 17s-T-put-PASS-FV 19.peanuts
     'On the table were put peanuts'

On the other hand, they are unlike DPs in several ways, and one of these is that when multiple locative expressions share a *v*P, a linker is optional:

(202) Omulume mo-a-sat-ire   omo-soko   (**m'**)
     1.man   AFF-1S/T-dance-EXT 18.LOC-market **18.LI**
     omo-nzoga (Kinande)
     18.LOC-bells
     'The man danced in the market with bells'

Baker and Collins suggest that locative expressions optionally sport a Case feature; equivalently, in this account, we can say that they may be treated either as PP or as DP.

The behavior of these expressions in the presence of an internal DP argument is complex, and I will be unable to fully account for it here. Baker and Collins (2006, 336–337) note, for example, that the linker is optional in (203a), but actually dispreferred in (203b):

(203) a. Omwami a-seny-er-aw-a   olukwi (**l'**)
        1-chief 1s/T-chop-APPL-PASS-FV 11-wood **11.LI**
        omo-mbasa (Kinande)
        18.LOC-axe
        'The chief was chopped wood with an axe'
     b. Olukwi lw-a-seny-er-aw-a   omwami (??**y'**) omo-mbasa
        11-wood 11s-T-chop-APPL-PASS-FV 1-chief   **1.LI** 18.LOC-axe
        'The wood was chopped for the chief with an axe'

If *omo-mbasa* 'with an axe' is capable of being treated as either a PP or a DP, we ought to expect the linker to be optional in both of these examples; if *omo-mbasa* is a PP, it can be safely linearized together with the other DP in the *v*P, and if it is a DP, it cannot be and a linker is needed. The fact that a linker is actually dispreferred in (203b) is unexpected.

Also unexpected are Pierre Mujomba's judgments of structurally similar sentences; for him, the linker is required in the equivalent of (203a), and fully optional in (203b):

(204) a. Ómwamy' a-seny-er-áw'   êsyóngwé *(**sy'**) omó-mbása ya
        1-chief   1s/T-chop-APPL-PASS 9-wood   **9.LI** 18.LOC-axe of
        Kámbale (Kinande)
        Kambale
        'The chief was chopped wood with Kambale's axe'
     b. Esyóngwé sy-a-seny-er-áw'   ómwamí (**y'**) omó-mbása
        9-wood   9s-T-chop-APPL-PASS 1-chief   **1.LI** 18.LOC-axe
        ya Kámbale
        of Kambale
        'The wood was chopped for the chief with Kambale's axe'

Finally, Baker and Collins note that in active sentences containing both a locative expression and a DP internal argument, the linker appears to be required:

(205) Kambale mo-a-teta-gul-a   eritunda *(**ry'**)
      Kambale AFF-1s-NEG.PAST-buy-FV 5.fruit   **5.LI**
      omo-soko (Kinande)
      18.LOC-market
      'Kambale didn't buy the fruit in the market'

Again, this is unexpected. In what follows I will abstract away from these facts, assuming that nominals prefixed with *omo-* can be treated as either PP or DP, but it is clear that there is more investigation to do; under circumstances that are still not clear, it seems that locative expressions are preferentially represented as DPs.[49]

The other class of expressions in Kinande that are of interest to us here are a type of narrow-scope indefinite. All of the DPs discussed so far begin with an initial vowel that Baker and Collins refer to as the "augment." Augmented and augmentless nominals receive different interpretations:

(206) a. Kambale mo-a-teta-gul-a   e̱-ri-tunda (Kinande)
      Kambale AFF-1S-NEG.PAST-buy-FV AUG-5-fruit
      'Kambale did not buy the/a certain fruit'
   b. Kambale mo-a-teta-gul-a   ri-tunda
      Kambale AFF-1S-NEG.PAST-buy-FV 5-fruit
      'Kambale did not buy a/any fruit'

Augmentless nominals also differ from ordinary DPs in that, given a choice between an augmentless nominal and an ordinary DP, it must be the ordinary DP that moves to prelinker position:

(207) a. Si-n-andisyata-hek-er-a   o-mu-kali   *(yo)
      NEG-1sgS-FUT-carry-APPL-FV AUG-1-woman 1.LI
      ka-tebe (Kinande)
      12-pail
      'I will not carry any pail for the woman'
   b. *Si-n-andisyata-hek-er-a   ka-tebe (k')   o-mu-kali
      NEG-1sgS-FUT-carry-APPL-FV 12-pail 12.LI AUG-1-woman

As Baker and Collins point out, augmentless nominals very generally avoid controlling agreement. For instance, an augmentless nominal cannot appear in the preverbal position, where it would control agreement on the verb (Pierre Mujomba, personal communication):

(208) *(A-)bá-kalí   ba-ká-seny' e-syó-ngwé (Kinande)
     AUG-2-women 2s-HAB-chop AUG-10-wood
     '(The) women chop the wood'

What happens when there are multiple augmentless nominals in the *v*P? Here there appears to be variation among speakers. Baker and Collins report that for Philip Ngessimo Mutaka, the Kinande speaker who worked with them, the linker cannot appear in such a sentence:

(209) Si-n-andisyata-hek-er-a   mu-kali (***yo**) ka-tebe (Kinande)
        NEG-1SG.S-FUT-carry-APPL-FV 1-woman **1.LI** 12-pail
        'I will not carry any pail for any woman'

For Pierre Mujomba (personal communication), by contrast, the linker
must appear in this kind of example:

(210) Si-n-andisyata-hek-er-a   mu-kali *(**yo**) ka-tebe (Kinande)
        NEG-1SG.S-FUT-carry-APPL-FV 1-woman **1.LI** 12-pail
        'I will not carry any pail for any woman'

Suppose we begin with Mujomba's judgments. If he must choose between placing an augmented DP and an augmentless DP before the
linker, he prefers the augmented DP, as we see in (207). If there are only
augmentless DPs in the *v*P, however, he will move one of them to the prelinker position; this is shown in (210). For Mujomba, then, augmented
and augmentless nominals are the same as far as Distinctness is concerned; if the *v*P contains two nominals (with or without augments), then
one must move to the prelinker position. His preference is to avoid put-
ting an augmentless nominal in a position where it would control agreement, such as the prelinker position, but he will do this if it is the only
way to rescue a structure from a Distinctness violation.

Philip Ngessimo Mutaka's judgments are more difficult to account for.
The difference between him and Mujomba has to do with sentences with
multiple postverbal augmentless DPs; for Mutaka, such sentences lack a
linker, while for Mujomba the linker is present. The easiest way I can
think of to accommodate Mutaka's judgments in the framework developed here is to say that his sentences are syntactically identical to
Mujomba's, but that his need to avoid agreement with an augmentless
nominal drives him to delete the linker when an augmentless nominal is
forced to move into the prelinker position.

There is at least one difference, for both speakers, between sentences
with multiple augmentless DPs after the verb and sentences with multiple
augmented DPs after the verb. Recall that, as we saw above, when the
DPs have their augments, they may appear in any order:

(211) a. Si-n-andisyata-hek-er'   o-mu-kali   **y'**
            NEG-1SG.S-FUT-carry-APPL AUG-1-woman 1.LI
            a-ka-tebe (Kinande)
            AUG-12-pail
            'I will not carry the pail for the woman'

b. Si-n-andisyata-hek-er' a-ka-tebe k' o-mu-kali
NEG-1SG.S-FUT-carry-APPL AUG-12-pail 12.LI AUG-1-woman
'I will not carry the pail for the woman'

For both speakers, it is impossible to alter the order of the augmentless nominals:

(212) a. Si-n-andisyata-hek-er-a   mu-kali (% yo)
NEG-1SG.S-FUT-carry-APPL-FV 1-woman 1.LI
ka-tebe (Kinande)
12-pail
'I will not carry any pail for any woman'
b. *Si-n-andisyata-hek-er-a   ka-tebe (ko) mu-kali
NEG-1SG.S-FUT-carry-APPL-FV 12-pail 12.LI 1-woman
'I will not carry any pail for any woman'

Baker and Collins suggest, on the basis of (211), that the Minimal Link Condition is suspended for this type of movement. I have no real theory to suggest about why this would be so just for augmented DPs. Given that augmentless DPs are nonspecific indefinites, we might try to link the facts in (211) and (212) to another contrast in Minimal Link Condition effects, namely the fact that Superiority effects apply to ordinary *wh*-phrases, but are suspended for D-linked *wh*-phrases (see Pesetsky 1987, 2000, and much other work):

(213) a. Which boy bought which book?
b. Which book did which boy buy?

(214) a. Who bought what?
b. *What did who buy?

To sum up what we have seen so far, in Kinande, when a *v*P contains two internal DP arguments, one must move to a higher position, characterized in Baker and Collins 2006 as the specifier of a functional head which is realized morphologically as the linker. In what follows we will consider what happens when the *v*P contains three different internal DP arguments.

Consider, first, the examples in (215), which have a direct object DP and a locative in the postverbal field (Pierre Mujomba, personal communication):

(215) a. U-ká-seny-er'   esyóngwé **sy'** omó músítu. (Kinande)
         2ss-HAB-chop-APPL 9-wood   9.LI 18-in 3-forest
         'You chop wood in the forest'
      b. U-ká-seny-er'   omó músítú **mw'** esyóngwé.
         2ss-HAB-chop-APPL 18-in 3-forest 18.LI 9-wood
         'You chop wood in the forest'

The facts in (215) are ones we have seen before; the postverbal field contains a linker, and either of the two postverbal phrases may move to its specifier. Next, let us add another postverbal DP (Pierre Mujomba, personal communication):[50]

(216) a. U-ká-seny-er'   ómwamí **y'** esyóngw' omó
         2ss-HAB-chop-APPL 1-chief 1.LI 9-wood   18-in
         músítu (Kinande)
         3-forest
         'You chop wood in the forest for the chief'
      b. U-ká-seny-er'   esyóngwé **sy'** ómwamy' omó músítu.
         2ss-HAB-chop-APPL 9-wood   9.LI 1-chief   18-in 3-forest
         'You chop wood in the forest for the chief'
      c. *U-ká-seny-er'   omó músítú **mw'** ómwamy' esyóngwé.
         2ss-HAB-chop-APPL 18-in 3-forest 18.LI 1-chief   9-wood
         'You chop wood in the forest for the chief'

As (216) demonstrates, in a sentence containing two postverbal DPs and a postverbal locative, either of the DPs may move to the prelinker position, but the locative may not.[51] This is not because the locative may not move to the prelinker position in general; we saw in (215b) above that it can. Rather, it has to do with the fact that (adapting a theory from Baker and Collins) the locative may be interpreted as either a DP or a PP. Consequently, when a VP contains two DPs and a locative, Distinctness requires us to move one of the DPs into the specifier of the linker's projection; if the locative is interpreted as a PP, the VP will be left with one DP and one PP, and will be linearizable. Example (216a) has the tree in (217); here I have omitted the subject, which I assume will move to the specifier of CP as before:

(217)

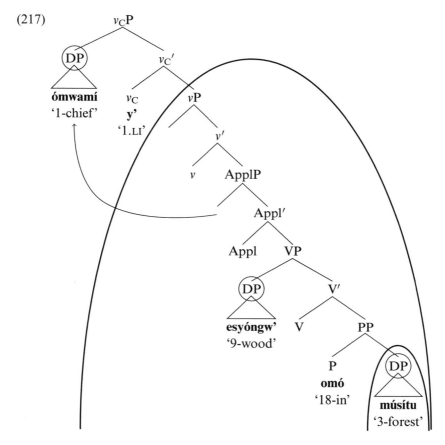

In the tree in (217), the overt DPs are all in separate Spell-Out domains.[52] *Ómwamí* '1-chief' is in the highest one, separated from the rest of the postverbal field by the Linker, which is (by hypothesis) the previously posited phase head v_C; *músítu* '3-forest' is in the lowest Spell-Out domain, protected from the rest of the VP by the preposition *omó*; and *esyóngwé* '9-wood' is in the middle domain. If we were instead to move *omó músítu* '18-in 3-forest' into the specifier of the v_CP, we would leave two DPs in the lowest Spell-Out domain, violating Distinctness:

(218)

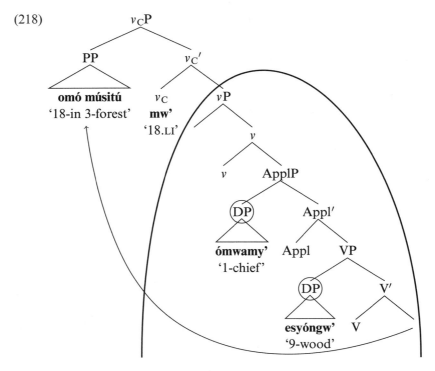

In (218), the lowest Spell-Out domain contains two DPs (*ómwamy'* '1-chief' and *esyóngwé* '9-wood') and is therefore unlinearizable.

In fact, there are several strategies for redeeming an ill-formed structure like (218). One is to *wh*-extract either of the offending DPs:[53]

(219) a. Ní ndi y'   ú-ká-seny-er'   omó músitú mw'
　　　 is who 1.LI 2SS-HAB-chop-APPL 18-in 3-forest 18.LI
　　　 esyóngwé? (Kinande)
　　　 9-wood
　　　 'For whom do you chop wood in the forest?'
　　 b. Ní ki   ky' ú-ká-seny-er'   omó músitú mw' ómwami?
　　　 is what 7.LI 2SS-HAB-chop-APPL 18-in 3-forest 18.LI 1-chief
　　　 'What do you chop in the forest for the chief?'

The examples in (219) have trees like the one in (218) as an intermediate stage; the locative expression *omó músitú* '18-in 3-forest' has been moved into the specifier of the linker's phrase, leaving two DPs inside the lowest Spell-Out domain. As long as one of these DPs is *wh*-extracted, however, the resulting structure is linearizable. This is not because of a general amnesty for *wh*-phrases; Kinande can optionally leave *wh* in situ, but the *wh* in situ equivalents of the examples in (219) are ill formed:

(220) *U-ká-seny-er'   omó músitú **mo** ndi esyóngwé? (Kinande)
2ss-HAB-chop-APPL 18-in 3-forest 18.LI who 9-wood
'For whom do you chop wood in the forest?'

Another way to rescue the structure in (218) is to delete one of the DPs, replacing it with an agreement morpheme in the verbal complex:

(221) a. U-ká-**mú**-seny-er'   omó músitú **mw'** esyóngwé (Kinande)
2ss-HAB-**1.O**-chop-APPL 18-in 3-forest 18.LI 9-wood
'You chop wood in the forest for **him**'
  b. U-ká-**sí**-seny-er'   omó músitú **mw'** ómwami
2ss-HAB-**9.O**-chop-APPL 18-in 3-forest 18.LI 1-chief
'You chop **it** in the forest for the chief'

In (221), the domain following the linker contains only one overt DP, and the examples are well formed.

If these examples involve pro-drop licensed by an agreement morpheme, then we will need to complicate our understanding of how phonologically null phrases are treated by linearization; the claim made elsewhere in this chapter (see the introduction and section 2.2.1.1) has been that the syntax is unaware of the phonology associated with particular functional heads, and therefore seeks to linearize them all. In the Kinande examples in (221), the phonologically null DPs appear to escape linearization, so we would need to allow the linearization system to be aware that certain kinds of things are null (perhaps because they are actually deleted via a syntactic operation).

Alternatively, we could understand the examples in (221) as involving cliticization. The Kinande verb is typically written as a single word, but this need not be the correct analysis. If there is a word break before the verb stem *seny-*'chop' in (221), for instance, then we could interpret the object agreement markers as proclitics attaching to the verb. Interestingly, there are phonological processes that group the stem together with the object agreement markers, to the exclusion of preceding material; work on Kinande and on Bantu more generally refers to the string beginning with the object marker as the "macrostem" (see Mutaka 1994, Black 1995, Archangeli and Pulleyblank 2002, as well as Jones and Coon 2008).[54] We might conclude from this that the verb stem does indeed begin a separate word in some sense, and that the macrostem is formed by procliticizing object clitics to that word. The preceding morphology would then have to be analyzed as an auxiliary or string of auxiliaries, perhaps also attached to the verb.[55]

Another way the Kinande VP can be made unacceptably full of DPs involves a causative construction. Kinande has morphological causatives:

(222) N-gá-seny-esay'   abákalí **b'** esyóngwé (Kinande)
      1ss-HAB-chop-CAUS 2-women 2.LI 9-wood
      'I make the women chop wood'

If we causativize a verb that has already been made ditransitive (by adding an applicative argument to a transitive verb), the resulting clause will have three VP-internal DP arguments:

(223) *N-gá-seny-es-er-ay-a   Kámbalé **y'** abákaly'
      1ss-HAB-chop-CAUS-APPL-CAUS-FV Kambale 1.LI 2-women
      esyóngwé (Kinande)
      9-wood
      'I make the women chop wood for Kambale' *or*
      'I make Kambale chop wood for the women'

This sentence is ill-formed on any meaning, nor can it be saved by re-ordering the postverbal DPs; to express these meanings, a biclausal construction must be used. The account developed here allows us to see why. Kinande has a strategy for rescuing a VP that contains two DPs; one can be moved to the specifier of the projection headed by the linker, thereby allowing the two DPs to be linearized in different Spell-Out domains. For a VP which has three DPs, this strategy will not work; only one DP can move to the specifier of the linker, and the VP will still have two DPs in it, rendering it unlinearizable.

As the previous section might lead us to expect, an example like (223) can be repaired by overtly *wh*-extracting one of the DPs:

(224) a. Ní ndi y' ú-ká-seny-es-er-ay'   abákalí **b'**
         is who 1.LI 2ss-HAB-chop-CAUS-APPL-CAUS 2-women 2.LI
         esyóngwé? (Kinande)
         9-wood
         'For whom do you make the women chop wood?' *or*
         'Who do you make chop wood for the women?'
      b. Ní ki   ky' ú-ká-seny-es-er-ay-a   Kámbalé
         is what 7.LI 2ss-HAB-chop-CAUS-APPL-CAUS-FV Kambale
         y' ómwami?
         1.LI 1-chief
         'What do you make Kambale chop for the chief?'

In (224), the postverbal field contains only two overt DPs, and thanks to the Linker's role as a phase head, they will be spelled out in separate Spell-Out domains, thereby obeying Distinctness.

We may descriptively summarize the Kinande facts so far as follows. The Kinande verb may be followed by at most two DPs; the examples discussed here have involved direct objects, benefactive applied objects, and causees. If there are two DPs, they must be separated by the linker. Methods for removing DPs from the postverbal field, we saw, include overt *wh*-extraction and pronominalization (which appears to be either pro-drop or cliticization).

I offered an analysis of these facts in terms of Distinctness. The linker is a phase head, which divides the postverbal field into two Spell-Out domains; consequently, if there are two DPs in the postverbal field, one must be in the specifier of the linker, where it can avoid being in the same Spell-Out domain with the other DP. We can link these facts to the English and Chinese facts discussed in the previous two sections; in Kinande, as in English and Chinese, when a VP contains two DPs, one of them must move to a higher specifier, putatively the specifier of a phase head.

Although I have discussed a variety of strategies for avoiding Distinctness violations in this chapter, I have had little to say about which strategy ought to be used in which context. To a certain extent, we have seen that this is a point of cross-linguistic variation; some languages have construct state as a way of expressing possession, for example, and others do not. In what follows we will consider the behavior of Distinctness in two languages in which Distinctness effects are particularly richly attested, namely Kinande and Japanese. We will see some evidence that Distinctness effects are to be dealt with as rapidly as possible, and that this constrains the range of possible responses to particular problems.

The preceding discussion of Kinande has focused entirely on relations between internal arguments. Once we consider the behavior of the Kinande subject, we discover a number of other phenomena that seem susceptible to explanation in terms of Distinctness.

As was mentioned earlier, Kinande has a process of locative inversion, whereby locative expressions move to the preverbal position and take over subject agreement (Pierre Mujomba, personal communication):

(225) a. Abákali ba-ká-seny-er'   omó músítu (Kinande)
  2-women 2s-HAB-chop-APPL 18-in 3-forest
  'The women chop in the forest'

b. Omó músitú mu-ká-seny-er'   abákali
   18-in 3-forest 18s-HAB-chop-APPL 2-women
   'In the forest, women chop'

Just as in English, however, this process cannot take place if the verb is
transitive (Pierre Mujomba, personal communication):

(226) a. Abákalí ba-ká-seny-er'   omó músitú mw'
         2-women 2s-HAB-chop-APPL 18-in 3-forest 18.LI
         esyóngwé (Kinande)
         9-wood
         'The women chop wood in the forest'
      b. *Omó músitú mu-ká-seny-er'   abákalí b' esyóngwé
         18-in 3-forest 18s-HAB-chop-APPL 2-women 2.LI 9-wood
      c. *Omó músitú mu-ká-seny-er'   esyóngwé sy' abákali
         18-in 3-forest 18s-HAB-chop-APPL 9-wood   9.LI 2-women

As (226) shows, when the verb has an object, locative inversion becomes
impossible, regardless of whether the postverbal subject precedes the object (as in (226b)) or follows it (as in (226a)).

We also saw earlier that Kinande has a process of Subject-Object Reversal (Pierre Mujomba, personal communication):

(227) a. Omúlumy' a-ámá-hek'   akatébé (Kinande)
         1-man   1s-PRES-carry 12-bucket
         'The man carries the bucket'
      b. Akatébé ka-ámá-hek-á   múlúme
         12-bucket 12s-PRES-carry-FV 1-man
         'The MAN carries the bucket'

Again, subject-verb inversion is impossible if the postverbal field contains
another DP (Pierre Mujomba, personal communication):

(228) a. Omúlumy' a-ámá-hek-er'   omúkalí y'
         1-man   1s-PRES-carry-APPL 1-woman 1.LI
         akatébé (Kinande)
         12-bucket
         'The man carries the bucket for the woman'
      b. *Akatébé ka-ámá-hek-er'   omúkalí (yo) múlúme
         12-bucket 12s-PRES-carry-APPL 1-woman 1.LI 1-man
         'The MAN carries the bucket for the woman'
      c. *Akatébé ka-ámá-hek-er-a   múlúme (y') omúkalí
         12-bucket 12s-PRES-carry-APPL 1-man   1.LI 1-woman
         'The MAN carries the bucket for the woman'

Finally, Kinande has a construction, which I will refer to as the "expletive construction," in which nothing at all appears in the preverbal position; the subject agreement morpheme takes on an invariant form, perhaps via agreement with a null expletive, and the subject is interpreted with focus (Halpert 2008; Miyagawa, forthcoming):

(229) a. Mo-ha-sat-ire   mukali. (Kinande)
        AFF-16s-dance-PAST 1-woman
        'A WOMAN danced'
    b. Mo-ha-gul-ir-we   ritunda
        AFF-16s-buy-PAST-PASS 5-fruit
        'FRUIT was bought'

Again, this construction is impossible with transitive verbs (Pierre Mujomba, personal communication):

(230) a. *Mó-há-gúl-iry' (o)-mú-kali   y' (e)-ri-túnda (Kinande)
        AFF-16s-buy-PAST AUG-1-woman 1.LI AUG-5-fruit
        'A WOMAN bought fruit'
    b. *Mó-há-gúl-iry' (e)-ri-túnda ry' (o)-mú-kali
        AFF-16s-buy-PAST AUG-5-fruit 5.LI AUG-1-woman
        'A WOMAN bought fruit'

If we ignore everything we have learned about Kinande, these look like good cases for Distinctness. In all three of the constructions named above, the subject is prevented from appearing in postverbal position just in case another DP appears there.

In the previous discussion of Kinande, however, we established that Kinande has a variety of methods for dealing with Distinctness violations involving multiple internal DP arguments. It turns out that none of these methods are effective when it comes to interactions between subjects and other DPs.

For instance, we saw that when the VP contains two DPs that are internal arguments, along with a PP, the two DPs must be separated by the linker (Pierre Mujomba, personal communication):

(231) a. U-ká-seny-er'   ómwamí y' esyóngw' omó
        2ss-HAB-chop-APPL 1-chief 1.LI 9-wood 18-in
        músítu (Kinande)
        3-forest
        'You chop wood in the forest for the chief'

b. U-ká-seny-er'   esyóngwé **sy'** ómwamy' omó músítu.
   2ss-HAB-chop-APPL 9-wood   9.LI 1-chief   18-in 3-forest
   'You chop wood in the forest for the chief'

c. *U-ká-seny-er'   omó músítú **mw'** ómwamy' esyóngwé.
   2ss-HAB-chop-APPL 18-in 3-forest 18.LI 1-chief   9-wood
   'You chop wood in the forest for the chief'

The account given above was that the linker is a phase head, and hence is capable of segregating the two DPs into separate Spell-Out domains.

As noted earlier, in the Kinande Subject-Object reversal construction, the subject must follow the linker (Pierre Mujomba, personal communication):[56]

(232) a. Esyóngwé si-ká-seny-er'   omó músítú mó
         9-wood   9s-HAB-chop-APPL 18-in 3-forest 18.LI
         bákali (Kinande)
         2-women
         'WOMEN chop wood in the forest'

      b. *Esyóngwé si-ká-seny-er'   bákalí   b'   omó músítú
         9-wood   9s-HAB-chop-APPL 2-women 18.LI 18-in 3-forest
         'WOMEN chop wood in the forest'

However, Subject-Object Inversion is impossible in the presence of another VP-internal DP argument, regardless of the order of the subject and the linker:

(233) a. Omúlumy' a-ámá-hek-er'   omúkalí y'
         1-man   1s-PRES-carry-APPL 1-woman 1.LI
         akatébé (Kinande)
         12-bucket
         'The man carries the bucket for the woman'

      b. *Akatébé ka-ámá-hek-er'   omúkalí (yo) múlúme
         12-bucket 12s-PRES-carry-APPL 1-woman 1.LI 1-man
         'The MAN carries the bucket for the woman'

      c. *Akatébé ka-ámá-hek-er-a   múlúme (y') omúkalí
         12-bucket 12s-PRES-carry-APPL 1-man   1.LI 1-woman
         'The MAN carries the bucket for the woman'

The ill-formedness of (233b,c) is surprising. If the linker can intervene between the two DPs, why should this not save the structure, as before?

Similarly, we saw above that a sentence with two VP-internal DP arguments may put a non-DP in prelinker position, as long as one of the DPs is replaced with an agreement morpheme in the verbal complex. Thus, the examples in (234a,b) contrast with the one in (234c):

(234) a. U-ká-**mú**-seny-er'   omó músitú **mw'**
2ss-HAB-**1o**-chop-APPL 18-in 3-forest 18.LI
esyóngwé. (Kinande)
9-wood
'You chop wood in the forest for **him**'

b. U-ká-**sí**-seny-er'   omó músitú **mw'** ómwami.
2ss-HAB-**9o**-chop-APPL 18-in 3-forest 18.LI 1-chief
'You chop **it** in the forest for the chief'

c. *U-ká-seny-er'   omó músitú **mw'** ómwamy' esyóngwé.
2ss-HAB-chop-APPL 18-in 3-forest 18.LI 1-chief   9-wood
'You chop wood in the forest for the chief'

The contrasts in (234) were offered as an argument for Distinctness. The problem in (234c) is not a problem with argument structure, or with the locality conditions on movement to the prelinker position; rather, it is an instance of a ban on two DPs in close structural proximity, part of the general system of such bans that we have been investigating in this chapter.

However, this process of pro-drop or cliticization cannot remedy Distinctness violations involving postverbal subjects. This is demonstrated for Subject-Object Reversal in (235), and for locative inversion in (236):

(235) a. *Akatébé   ka-ámá-hek-er'   omúkalí (yo)
12-bucket 12s-PRES-carry-APPL 1-woman 1.LI
múlúme (Kinande)
1-man
'The MAN carries the bucket for the woman'

b. *Akatébé   ka-ámá-**mu**-hek-er-á   múlúme
12-bucket 12s-PRES-**1o**-carry-APPL-FV 1-man
'The MAN carries the bucket for **her**'

(236) a. *Omó músitú   mu-ká-seny-er'   abákalí b' esyóngwé
18-in 3-forest 18s-HAB-chop-APPL 2-women 2.LI 9-wood
'In the forest, the women chop wood'

b. *Omó músitú   mu-ká-**sí**-seny-er'   abákali
18-in 3-forest 18s-HAB-**9o**-chop-APPL 2-women
'In the forest, the women chop it'

Similarly, we saw earlier that violations of Distinctness involving multiple internal DP arguments can be remedied by *wh*-extraction of one of the DPs:

(237) a. Ní ndi y' ú-ká-seny-es-er-ay'   abákalí **b'**
           is who 1.LI 2SS-HAB-chop-CAUS-APPL-CAUS 2-women 2.LI
           esyóngwé? (Kinande)
           9-wood
           'For whom do you make the women chop wood?' *or*
           'Who do you make chop wood for the women?'
      b. Ní ki   ky' ú-ká-seny-es-er-ay-a   Kámbalé **y'**
           is what 7.LI 2SS-HAB-chop-CAUS-APPL-CAUS-FV Kambale 1.LI
           ómwami?
           1-chief
           'What do you make Kambale chop for the chief?'
      c. *N-gá-seny-es-er-ay-a   Kámbalé **y'** abákaly'
           1SS-HAB-chop-CAUS-APPL-CAUS-FV Kambale 1.LI 2-women
           esyóngwé
           9-wood
           'I make the women chop wood for Kambale' *or*
           'I make Kambale chop wood for the women'

And again, we find that when the Distinctness violation involves a post-verbal subject, *wh*-extraction cannot save the structure. This is demonstrated for the expletive construction in (238), and for locative inversion in (239):

(238) a. *Mó-há-gúl-iry' (o)-mú-kali   y' (e)-ri-túnda (Kinande)
           AFF-16S-buy-PAST AUG-1-woman 1.LI AUG-5-fruit
           'A WOMAN bought fruit'
      b. *Ní ki   kyo mó-há-gúl-iry' (o-)mú-kali?
           is what 7.LI AFF-16S-buy-PAST AUG-1-woman
           'What did a WOMAN buy?'

(239) a. *Omó músitú mu-ká-seny-er'   abákalí b' esyóngwé
           18-in 3-forest 18S-HAB-chop-APPL 2-women 2.LI 9-wood
           'In the forest, the women chop wood'
      b. *Ní ki   ky' omó músitú mu-ká-seny-er'   abákali?
           is what 7.LI 18-in 3-forest 18S-HAB-chop-APPL 2-women
           'What, in the forest, do women chop?'

There are still many questions to answer about Kinande syntax, and it could be that some of the facts in this section will receive independent explanations. Perhaps the facts about *wh*-extraction in (238) and (239), for example, have to do with the interaction between *wh*-movement and focus, and are unrelated to Distinctness. Similarly, we might discover facts about the placement of the linker in the structure of *v*P that explain the fact that linkers cannot save Distinctness violations involving postverbal subjects.

At our current level of understanding, however, the data seem to point to a generalization: Distinctness violations involving postverbal subjects cannot be remedied in the same ways that Distinctness violations involving multiple internal arguments can be. Why should this be?

Of course, one possible response to these facts would be to retreat from the claim that all of these effects are Distinctness effects. On that view, we could take the option of rescue via *wh*-extraction of an offending DP, for example, as a crucial hallmark of Distinctness, and phenomena that lack this hallmark would require some other explanation. Later, however, I will develop an alternative account, intended to explain why some Distinctness violations are more difficult to remedy than others. Before we do this, however, let us turn to the behavior of a certain type of Distinctness effect in Japanese. We will see that in Japanese, as in Kinande, some violations of Distinctness are easier to repair than others.

The Double-o Constraint in Japanese (see Harada 1973; Kuroda 1988; Saito 2002; Poser 2002; Hiraiwa, forthcoming; and the references cited there) is another good candidate for a Distinctness effect. The Double-o Constraint bans instances of two structurally adjacent DPs that are both marked with the accusative suffix -*o*.

We can see one instance of the constraint in (240):

(240) a. Hanako-ga   Taroo-ni   toti-**o**   zyooto sita. (Japanese)
          Hanako-NOM Taroo-DAT land-ACC giving did
          'Hanako gave Taroo a piece of land'
      b. Hanako-ga   Taroo-ni   toti-no   zyooto-**o** sita.
          Hanako-NOM Taroo-DAT land-GEN giving-ACC did
      c. *Hanako-ga   Taroo-ni   toti-**o**   zyooto-**o** sita.
          Hanako-NOM Taroo-DAT land-ACC giving-ACC did

The examples in (240) involve the light verb construction *zyooto suru* 'do giving'. Either the theme DP *toti* 'land' or the noun *zyooto* 'giving' may be marked accusative (as in (240a,b)), but they cannot both be marked accusative at once (as in (240c)).

Another instance of the Double-o Constraint has to do with causatives. Japanese causatives are much like those discussed in section 2.1.3; causees of transitive predicates must be marked dative, while causees of intransitive predicates can be marked accusative (though they can also be marked Dative, unusually for this pattern) (Saito 2002):

(241) a. Hanako-ga   Taroo-**ni** hon-o   yom-aseru (Japanese)
          Hanako-NOM Taroo-DAT book-ACC read-CAUS
          'Hanako makes Taroo read a book'
      b. *Hanako-ga   Taroo-**o**   hon-o   yom-aseru
          Hanako-NOM Taroo-ACC book-ACC read-CAUS
          'Hanako makes Taroo read a book'

(242) Hanako-ga   Taroo-**ni/o**   hasir-aseru
      Hanako-NOM Taroo-DAT/ACC run-CAUS
      'Hanako makes Taroo run'

We can view the ill-formedness of (241b) as an instance of the Double-o Constraint; the causee *Taroo* cannot be marked accusative, because the direct object of the caused predicate, *hon* 'book', is already marked accusative.

Much work on the Double-o Constraint has uncovered two kinds of violations. Examples like the one in (240c), repeated below as (243a), are comparatively weakly ungrammatical. Moreover, they may be rendered completely grammatical by separating the two *o*-marked DPs from each other, for instance, via a cleft, as in (243b) (Saito 2002):

(243) a. *Hanako-ga   Taroo-ni toti-**o**   zyooto-**o**   sita. (Japanese)
          Hanako-NOM Taroo-DAT land-ACC giving-ACC did
          'Hanako gave Taroo a piece of land'
      b. [Hanako-ga   Taroo-ni zyooto-**o**   sita no wa] toti-**o**   da.
          Hanako-NOM Taroo-DAT giving-ACC did C TOP land-ACC is
          'What Hanako gave to Taroo is a piece of land'

By contrast, the Double-o Constraint violation involving a causative, in (241b) above, is more ungrammatical than (243a), and cannot be repaired via clefting:

(244) a. *Hanako-ga   Taroo-**o**   hon-o   yom-aseru (Japanese)
          Hanako-NOM Taroo-ACC book-ACC read-CAUS
          'Hanako makes Taroo read a book'
      b. *[Hanako-ga   Taroo-**o**   yomaseta   no wa] hon-**o**   da
          Hanako-NOM Taroo-ACC read-CAUS-PAST C TOP book-ACC is
          'What Hanako made Taroo read is a book'

c. *[Hanako-ga hon-**o** yomaseta no wa] Taroo-**o** da
Hanako-NOM book-ACC read-CAUS-PAST C TOP Taroo-ACC is
'The one that Hanako made read a book is Taroo'

We can construct causative examples with the properties of the "weaker" Double-o effect that we saw in the light-verb case, by using accusative-marked DPs that are not arguments. The DP *hamabe-o* 'beach-ACC' in (245) is one such DP (Saito 2002):

(245) Taroo-ga hamabe-o hasiru (Japanese)
Taroo-NOM beach-ACC run
'Taroo runs on the beach'

In the causativized version of (245), dative marking is preferred to accusative marking on the causee, but the ungrammaticality is the weak kind found with the light-verb construction in (243), not the strong kind found with the ordinary causative in (244). Moreover, just as with the light-verb construction, the Double-o constraint violation can be completely repaired via clefting:

(246) a. Hanako-ga Taroo-**ni**/*-**o** hamabe-o hasiraseru (Japanese)
Hanako-NOM Taroo-DAT/ACC beach-ACC run-CAUS
'Hanako makes Taroo run on the beach'
b. [Hanako-ga Taroo-**o** hasiraseta no wa] hamabe-**o** da
Hanako-NOM Taroo-ACC run-CAUS-PAST C TOP beach-ACC is
'What Hanako made Taroo run on is a beach'
c. [Hanako-ga hamabe-**o** hasiraseta no wa] Taroo-**o** da
Hanako-NOM beach-ACC run-CAUS-PAST C TOP Taroo-ACC is
'The one that Hanako made run on a beach is Taroo'

Just like Kinande, then, Japanese has two distinguishable types of Distinctness effects; one can be rescued by A-bar movement, and the other cannot.

The different types of Distinctness violations in Kinande and Japanese are exemplified in (247) and (248). The (a) examples below are instances of "strong Distinctness," which cannot be remedied by movement operations; the (b) examples show "weak Distinctness," which can be repaired:

(247) a. *Akatébé ka-ámá-hek-er' omúkalí (yo)
12-bucket 12s-PRES-carry-APPL 1-woman 1.LI
múlúme (Kinande)
1-man
'The MAN carries the bucket for the woman'

b. *U-ká-seny-er'   omó músitú mw' ómwamy' esyóngwé.
   2ss-HAB-chop-APPL 18-in 3-forest 18.LI 1-chief   9-wood
   'You chop wood in the forest for the chief'

(248) a. *Hanako-ga   Taroo-**o**   hon-**o**   yom-aseru (Japanese)
        Hanako-NOM Taroo-ACC book-ACC read-CAUS
        'Hanako makes Taroo read a book'
     b. *Hanako-ga   Taroo-ni   toti-**o**   zyooto-**o**   sita.
        Hanako-NOM Taroo-DAT land-ACC giving-ACC did
        'Hanako gave Taroo a piece of land'

Here I will offer an account of the contrasts in (247) and (248).

One crucial component of the account is a version of the intuition that was useful in the discussion of Spanish Distinctness effects in section 2.4.3: Distinctness-violating configurations are to be avoided as much as possible, not just as parts of final PF representations but throughout the derivation. We can express this intuition more or less formally as in (249):

(249) *Derivational Distinctness*
     Given a choice between operations, prefer the operation (if any) that causes a Distinctness violation to appear as briefly as possible in the derivation.

In the discussion of Spanish in section 2.4.3, we considered the limiting case of the principle in (249); given a choice between a derivation in which a Distinctness violation appears and one in which it does not appear, the grammar prefers the derivation in which it does not appear (even if both derivations, in the end, yield representations that obey Distinctness). In Kinande, we will see evidence that the grammar makes a further kind of distinction: if Distinctness violations are unavoidable, the grammar seeks to eliminate them as quickly as possible, and given a choice of which Distinctness violation to eliminate, the grammar chooses the one created most recently.

We will see that the Derivational Distinctness condition in (249) often overlaps with Shortest Attract in its effects, but that the two conditions are not identical. It seems reasonable to hope, however, that (249) and Shortest Attract could be made to follow from a single overarching constraint. If we think, for example, that Shortest Attract involves a requirement that heads Agree first with the most recently Merged instance of a phrase bearing the uninterpretable feature that Agree will check,[57] then we could see (249) and Shortest Attract as particular cases of (250), a version of Chomsky's (1995) Featural Cyclicity (to use the terminology of Richards 1999, 2001):

(250) Given a choice between operations, choose the operation (if any) that causes an uninterpretable element to appear as briefly as possible in the derivation (where uninterpretable elements include unintepretable features and Distinctness violations).

Less formally, the condition in (250) requires the grammar to deal first with the most recently created problem, whether this is an uninterpretable feature or a Distinctness-violating configuration.

Let us begin by considering the contrasts in (251) and (252):

(251) a. Omúlumy' a-ámá-hek' akatébé (Kinande)
          1-man   1s-PRES-carry 12-bucket
          'The man carries the bucket'
      b. Akatébé ka-ámá-hek-á   múlúme
          12-bucket 12s-PRES-carry-FV man
          'The MAN carries the bucket'

(252) a. Omúlumy' a-ámá-hek-er'   omúkalí y'
          1-man   1s-PRES-carry-APPL 1-woman 1.LI
          akatébé (Kinande)
          12-bucket
          'The man carries the bucket for the woman'
      b. *Akatébé ka-ámá-hek-er'   omúkalí (yo) múlúme
          12-bucket 12s-PRES-carry-APPL 1-woman 1.LI 1-man
          'The MAN carries the bucket for the woman'

In (251), we see that Subject-Object Reversal is possible in Kinande transitive sentences; in (252), we see that it is no longer possible when an applicative object is added. Subject-Object Reversal requires the subject to lose its initial vowel augment (thus, in (251b), 'man' appears as *múlúme* rather than *omúlúme*). Since we saw earlier that nouns without augments are dispreferred as controllers of agreement in Kinande, we might take the loss of the subject's augment in (251b) as a way of circumventing conditions on locality of movement; perhaps the subject must be augmentless so that it does not act as a closer possible Goal for the Probe responsible for moving DPs out of the vP. Still, the facts in (252) show that while loss of the augment on the subject is necessary for Subject-Object inversion, it is not sufficient; even if the subject is augmentless, the object cannot invert if the clause also contains an applied object.

Consider first the derivation of a sentence like the one in (228a), repeated as (253):

(253) Omúlumy' a-ámá-hek-er'   omúkalí y'   akatébé (Kinande)
      1-man   1s-PRES-carry-APPL 1-woman 1.LI 12-bucket
      'The man carries the bucket for the woman'

The derivation for (253) begins at the bottom of the tree, assembling the
verb phrase together with its direct object and applicative object:

(254)

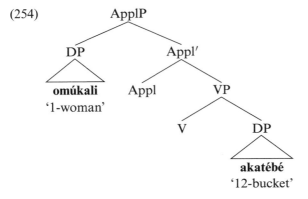

At this point, the tree is in violation of Distinctness. In Spanish, as we
saw in section 2.4.3, this Distinctness violation would be avoided by
Merging one of the DPs as a KP. Since the Kinande lexicon apparently
lacks K, Kinande does not have this option, and the Distinctness violation in (254) must simply be tolerated for the time being.

   The derivation of the sentence proceeds, Merging *v* and adding the external argument:

(255)

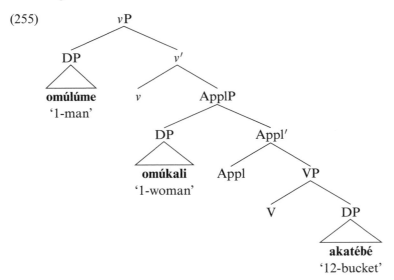

Adding the external argument creates new Distinctness violations, which must be repaired before the structure can be linearized; the new linearization statements that relate the subject to the object and the subject to the applied object will both be uninterpretable. Finally, the phase head v_C is Merged:

(256)

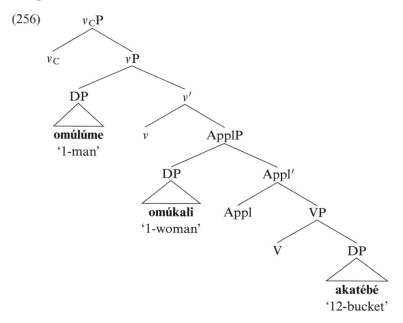

The Derivational Distinctness condition in (249) requires v_C to fix a Distinctness violation if it can. More specifically, (249) requires v_C to repair the Distinctness violation that was created most recently; this is the operation that will cause a Distinctness violation to have the shortest possible derivational lifespan. In this case, then, v_C is compelled to move the subject into its specifier, thereby eliminating the most recently created Distinctness violations:

(257)

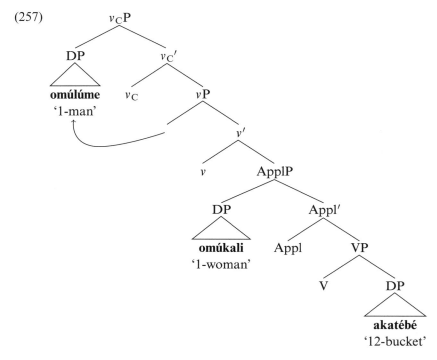

Since v_C is a phase head, movement of *omúlúme* '1-man' to the specifier of v_C eliminates the most recently created Distinctness violations, ensuring that the subject will be linearized in a different Spell-Out domain from the two internal arguments. Because the newest Distinctness violations both involve the subject, movement of the subject is the only single operation which will eliminate them all; movement of the direct object, for example, would leave untouched the linearization statement ordering the subject with the applied object. In this instance, Derivational Distinctness has the same effect that Shortest Attract would have; v_C is compelled to Agree with the closest available DP. We will later see a case in which Shortest Attract would fail to make the correct prediction.

Even after movement of the subject, the *v*P in (257) still contains a Distinctness violation, involving the two internal arguments. Since v_C has already eliminated the most recently created Distinctness violations, it is now free to turn its attention to this older violation:[58]

(258)

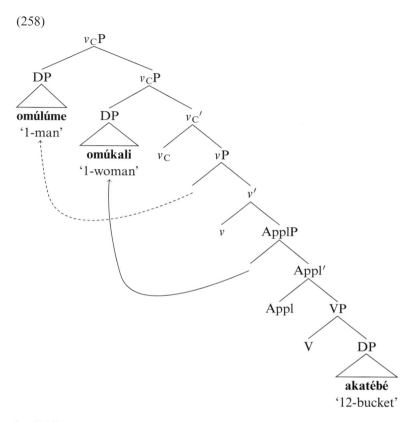

In (258), v_C Agrees with the applied object *omúkali* '1-woman', which moves into the specifier of v_C, tucking in underneath the existing specifier created in (257). The operation in (258) eliminates the first Distinctness violation that was created back in (254); the two internal arguments are now linearizable. Unfortunately, this step also creates a new violation; the two DPs in the specifiers of v_C are no longer linearizable. I assume that the grammar is insufficiently "intelligent" to notice these kinds of consequences of its actions; it performs operations that eliminate existing problems, but does not consider possible problems that these operations might cause.

The phase head v_C undergoes Spell-Out, sending its complement to PF; since this complement contains only one instance of DP, it is linearizable. The tree then continues to grow, adding (at least) T and C, and the subject moves into the specifier of CP (possibly stopping in the specifier of TP along the way):

(259)

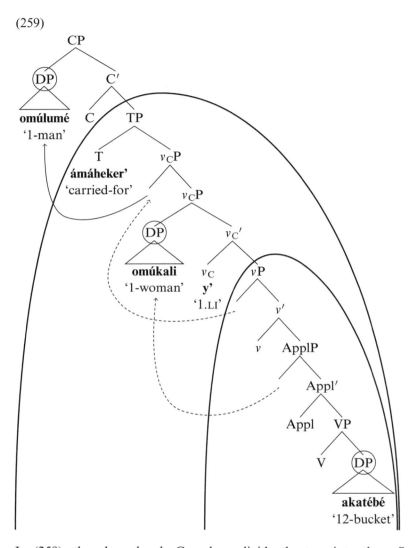

In (259), the phase heads C and v_C divide the tree into three Spell-Out domains, each of which contains one DP; Distinctness is therefore satisfied.

If C can only Agree with the closest available DP, movement of the subject into the specifier of CP is forced in (259); the applicative object *omúkali* '1-woman' is unable to move past the subject. Moreover, in a ditransitive sentence like this one, we arrive at the conclusion that only the subject can be in preverbal position. The Derivational Distinctness condition in (249), which requires the grammar to deal with the most recently created Distinctness violation first, forces v_C to raise the subject into its highest specifier; on the assumption that C can only Agree with the closest DP, this will mean that only the subject may raise into the specifier of CP.

We thus derive the ban on Subject-Object Inversion in sentences with multiple internal DP arguments:

(260) a. Omúlumy' a-ámá-hek-er'   omúkalí y'
       1-man   1s-PRES-carry-APPL 1-woman 1.LI
      akatébé (Kinande)
      12-bucket
      'The man carries the bucket for the woman'
   b. *Akatébé ka-ámá-hek-er'   omúkalí (yo) múlúme
      12-bucket 12s-PRES-carry-APPL 1-woman 1.LI 1-man
      'The MAN carries the bucket for the woman'

Both of the final representations in (260), on the theory presented above, obey Distinctness, in that they contain three DPs in three separate Spell-Out domains. Derivational Distinctness, however, imposes constraints on the derivation, requiring it to repair the most recently created Distinctness violation first. Consequently, the subject will be required to escape the vP first, and general conditions on locality of movement will then guarantee that the subject will end the derivation in the preverbal position. In (260b), therefore, although the final representation obeys Distinctness, the Distinctness violations were repaired in an improper order, permitting the Distinctness violations involving the subject to "survive" longer than Derivational Distinctness allows.

By contrast, consider the behavior of a transitive sentence:

(261) a. Omúlumy' a-ámá-hek'   akatébé (Kinande)
      1-man   1s-PRES-carry 12-bucket
      'The man carries the bucket'
   b. Akatébé ka-ámá-hek-á   múlúme
      12-bucket 12s-PRES-carry-FV man
      'The MAN carries the bucket'

In (261), either the subject or the object may be in preverbal position at the end of the derivation. This is as we expect. The derivation of (261) will begin with the creation of a Distinctness violation:

(262)

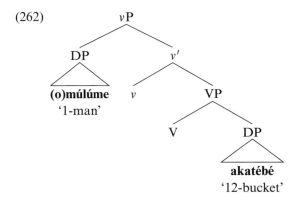

Merging the subject creates a Distinctness violation, which must be remedied. Crucially, however, the tree in (262) contains only one Distinctness violation, which may be repaired by movement of either DP to the specifier of v_CP:

(263) a.

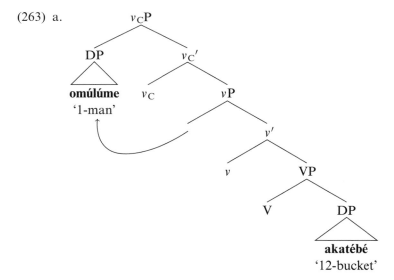

b.

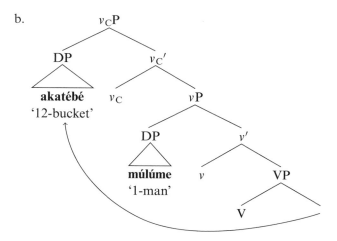

Either of the moves in (263) will eliminate the Distinctness violation created in (262), and either is therefore acceptable (though movement of the object past the subject requires the subject to drop its augment, perhaps for reasons having to do with locality, as suggested above). The DP that moves into the specifier of v_C will subsequently move into preverbal position, yielding the sentences in (264):

(264) a. Omúlumy' a-ámá-hek' akatébé (Kinande)
   1-man   1s-PRES-carry 12-bucket
   'The man carries the bucket'
   b. Akatébé ka-ámá-hek-á   múlúme
   12-bucket 12s-PRES-carry-FV man
   'The MAN carries the bucket'

Now we can return to the problem with which this section started. We have seen evidence for two types of Distinctness in Kinande. "Strong Distinctness" effects, exemplified in (265a), cannot be repaired by movement operations, while "weak Distinctness" effects, like the one in (265b), are susceptible to repair:

(265) a. *Akatébé ka-ámá-hek-er'   omúkalí (yo)
      12-bucket 12s-PRES-carry-APPL 1-woman 1.LI
      múlúme (Kinande)
      1-man
      'The MAN carries the bucket for the woman'
   b. *U-ká-seny-er'   omó músitú mw' ómwamy' esyóngwé.
      2ss-HAB-chop-APPL 18-in 3-forest 18.LI 1-chief   9-wood
      'You chop wood in the forest for the chief'

The discussion in this section has established a different way of viewing examples like (265a). In fact, the final representation in (265a) obeys Distinctness, in that the three DPs the clause contains are in separate Spell-Out domains. However, the Distinctness effects were repaired in an improper order. Once the phase head v_C gave the derivation a chance to begin alleviating Distinctness effects, it should have begun by moving the subject, which would have eliminated the most recently created violations, thereby obeying Derivational Distinctness. Instead, the derivation of (265a) chose to rectify an older Distinctness violation.

In (265b), by contrast, Distinctness is violated in the final representation; the Spell-Out domain represented by the postlinker field contains two DPs. We can take advantage of this contrast between (265a) and (265b) in accounting for the fact that (265b), but not (265a), may be repaired by removing one of the offending DPs:

(266) a. *Akatébé   ka-ámá-**mu**-hek-er-á   múlúme (Kinande)
           12-bucket 12s-PRES-**1o**-carry-APPL-FV 1-man
           'The MAN carries the bucket for **her**'
       b. U-ká-**mú**-seny-er'   omó músitú mw' esyóngwé.
           2ss-HAB-**1o**-chop-APPL 18-in 3-forest 18.LI 9-wood
           'You chop wood in the forest for **him**'

Converting the applied object into an agreement marker (or clitic) repairs (265b), but not (265a).

We are now in a position to see why this is so. Example (265a) violates a condition that effectively demands that in sentences containing multiple internal DP arguments, the subject must be made preverbal, thereby eliminating the Distinctness violations which are created last. This condition is violated both in (265a) and in (266a); in both examples, there are two vP-internal DPs, and the subject is postverbal. The same condition therefore rules out both examples; even if cliticization repairs the Distinctness violation involving the subject and the applied object, it does so too late to rescue the sentence. Fronting of the subject eliminates all the Distinctness violations created by Merge of the subject in a single operation; movement and cliticization of the two internal arguments accomplishes the same goal, but uses two operations to do so, violating the requirement that Distinctness violations be eliminated as quickly as possible.

In (266b), by contrast, there is no postverbal subject. Consequently, (266b) represents a repair of the only offending property of (265b); in (266b), unlike (265b), there is only one DP in the postlinker field. The well-formedness of (266b), and the contrast between "weak" and "strong" Distinctness effects, therefore follows.

Can this account be extended to the Japanese examples in (248), repeated here as (267)?

(267) a. *Hanako-ga   Taroo-**o**   hon-**o**   yom-aseru (Japanese)
          Hanako-NOM Taroo-ACC book-ACC read-CAUS
          'Hanako makes Taroo read a book'
      b. *Hanako-ga   Taroo-ni   toti-**o**   zyooto-**o**   sita.
          Hanako-NOM Taroo-DAT land-ACC giving-ACC did
          'Hanako gave Taroo a piece of land'

I believe that the Japanese, Kinande, and Spanish facts can be captured by the same principles, though a full account will entail a deeper understanding of the mechanisms of Case assignment in Japanese than I currently have.[59] Recall that (267b) is the example that can be repaired by clefting one of the *o*-marked DPs, while (267a) is not susceptible to repair in this way. If the account sketched above for Kinande is to apply here, then we will have to find some sense in which the Distinctness violation in (267a) could have been avoided or repaired earlier in the derivation, while the one in (267b) could not have been.

In the case of (267a), the way to avoid the Distinctness violation would be to mark the causee *Taroo* dative rather than accusative:

(268) Hanako-ga   Taroo-**ni**   hon-**o**   yom-aseru (Japanese)
      Hanako-NOM Taroo-DAT book-ACC read-CAUS
      'Hanako makes Taroo read a book'

The behavior of (267a) is thus very reminiscent of some of the Spanish facts discussed in section 2.4.3:

(269) a. Juan conoció *(**a**) Maria en Buenos Aires
          Juan met   a Maria in Buenos Aires
          'Juan met Maria in Buenos Aires'
      b. *(**A**) quién conoció Juan en Buenos Aires?
          a who met   Juan in Buenos Aires
          'Who did Juan meet in Buenos Aires?'

In (269a), *Maria* must be marked with *a*, for reasons that by now are familiar; by Merging *Maria* as a KP, we make it possible to linearize *Maria* with the subject DP (or ɸP) *Juan*. In (269b), the object *wh*-phrase *quién* must also be Merged as a KP, even though subsequent *wh*-movement

will leave the subject and the object in separate Spell-Out domains. The
conclusion offered in section 2.4.3 was that the grammar is not satisfied
with simply avoiding Distinctness in the final representation; Distinctness
violations must be avoided, as much as possible, at all times.

We can apply the same reasoning to the Japanese facts in (270a,b):

(270) a. *Hanako-ga   Taroo-**o**   hon-**o**   yom-aseru (Japanese)
         Hanako-NOM Taroo-ACC book-ACC read-CAUS
         'Hanako makes Taroo read a book'
   b. *[Hanako-ga   Taroo-**o**   yomaseta   no wa] hon-**o**   da
         Hanako-NOM Taroo-ACC read-CAUS-PAST C TOP book-ACC is
         'What Hanako made Taroo read is a book'

The Japanese example in (270b) has a derivation that first creates a Distinctness violation, then repairs it by separating the two accusative DPs
from each other via clefting. Since the initial Distinctness violation was
avoidable, its subsequent repair is insufficient to make the sentence well
formed; a derivation that marks the causee dative avoids the creation of
the Distinctness violation in the first place, and is therefore preferable to
the derivation that yields (270b).

As for (267b) (repeated as (271a)), there are in fact different ways of
assigning Case that make this example well formed:

(271) a. *Hanako-ga   Taroo-ni   toti-**o**   zyooto-**o**   sita. (Japanese)
         Hanako-NOM Taroo-DAT land-ACC giving-ACC did
         'Hanako gave Taroo a piece of land'
   b. Hanako-ga   Taroo-ni   toti-**no**   zyooto-**o**   sita.
         Hanako-NOM Taroo-DAT land-GEN giving-ACC did
   c. Hanako-ga   Taroo-ni   toti-**o**   zyooto sita.
         Hanako-NOM Taroo-DAT land-ACC giving did

If we are to characterize (271a) as an "unavoidable" violation of Distinctness (hence one that is repairable by such operations as clefting), we will
have to derive (271a) without considering the options in (271b,c). For
instance, we might pursue a theory in which the examples in (271) all involve different sets of functional heads; perhaps *toti* 'land' can only be
marked genitive if *zyooto* 'giving' has a certain amount of nominal structure which is absent in (271a), for example. If the reference set of options
to be considered as the derivation proceeds is at least partly determined
by the contents of the Numeration, then giving the examples in (239) different Numerations might lead us to the correct conclusion, namely that

the Distinctness violation in (271a) is unavoidable (given the particular Numeration from which (271a) is constructed), and can therefore be repaired by subsequent operations.

The discussion for most of this chapter has been concerned with trying to show that Distinctness effects exist, and with cataloging various ways of avoiding them. The harder work of showing how the grammar chooses among tactics for avoiding Distinctness violations will surely involve the kind of investigation begun here; we will need to find languages like Japanese and Kinande, in which Distinctness effects are richly attested, and try to understand how Distinctness-avoiding techniques are distributed in the grammar.

2.5 Case as Well as Case Resistance

In the course of this chapter we have seen Distinctness-based accounts of a number of the effects that are standardly attributed to Case. One difference between Distinctness and classic Case theory is that in the Distinctness approach, DPs are generally assumed to be licensed, unless they are brought unacceptably close to other DPs. In Case theory, by contrast, the default state of a DP is for it not to be licensed, unless it is sufficiently close to a Case-licenser.

We might be able to replace all or part of Case theory with Distinctness. This would not necessarily mean jettisoning the mechanisms of Case assignment; in fact, we have already used K(ase) heads to avoid Distinctness violations. Rather, we would be doing away with the stipulation that all DPs need Case, and replacing it with the more general condition of Distinctness. The new theory would be consistent with some nominals never receiving Case at all, or with Case-assignment structures being present to different extents in different languages, as long as Distinctness is respected.

In this section we will review the Distinctness-based accounts of certain Case phenomena, and add a few more.

2.5.1 Techniques for Satisfying Distinctness

In the approach developed here, Caselike phenomena have to do with interactions between DPs. We have seen two types of examples in which Distinctness effects may appear: one in which one DP dominates another, and another in which one DP asymmetrically c-commands another within a Spell-Out domain:

(272) a.

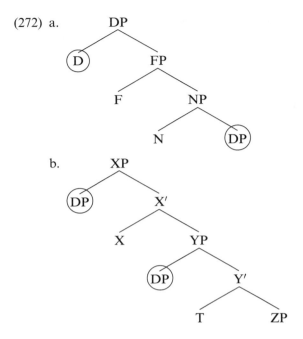

b.

2.5.1.1 DP Dominating DP Concentrating first on the dominance case in (272a), we have seen two kinds of ways of avoiding Distinctness violations. We can add material to one of the DPs, making it a KP or PP, or we can remove the functional material from one of the DPs, making it an NP. In fact, this second technique can be applied either to the dominated DP (as in (273b)) or to the dominating one (as in (273b′)):

(273) a.

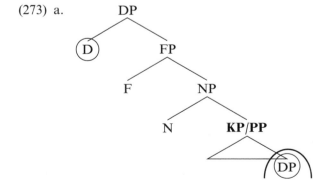

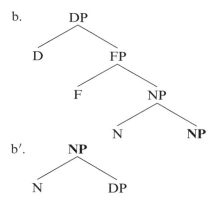

b.

b'.

The strategy in (273a) was the one we used to account for classic Case effects like the ones in (274):

(274) a. the destruction *(of) the city
    b. su amor *(al) dinero (Spanish)
     his love **a**-the money

Rather than simply stipulating that nouns are unable to assign Case, we can account for the facts in (274) by invoking the general principle of Distinctness. We have also seen the strategies in (273b) and (273b′) attested; these were, respectively, Hungarian nominative possessors and construct state in languages like Hebrew and Irish.

Part of the point of Distinctness is that these Case-driven phenomena are special instances of a general pattern. The strategies in (273) for linearizing structures with one DP dominated by another are used for possessors and noun complements, but also for relative-clause operators, as we have seen. (273a) is the strategy discussed in section 2.2.2.2, in which relative operators are required to be PP rather than DP;[60] (273b′), as we saw in section 2.4.2.2, is also attested, in languages like Akkadian that have the option of using construct state to license relative clauses. We might also attribute to strategy (273a) the use of genitive subjects in relative clauses in many languages:

(275) a. hu-me [**em** bič-ka-ʼu-m]
     the-PL **2.SGGEN** see-PERF-REL-PL
     sahak (Hiaki)
     leave-PL.PERF (Krause 2001, 43)
     'The ones who you saw left'

b. [**mini** aw-sen] mer^y-min^y (Dagur)
   I.GEN buy-PAST horse-1SG (Hale 2002, 109)
   'the horse that I bought'
c. [**John-no** yonda] hon (Japanese)
   John-GEN read book
   'the book that John read'

Krause (2001) argues that relative clauses of this type are invariably reduced relatives, lacking a CP layer and possibly some other functional structure as well. If she is correct, then the subject of such a reduced relative will be in the same Spell-Out domain as the functional material of the head DP. I will avoid here assigning a particular label to the functional structure of the reduced relative, labelling it RelP:

(276)

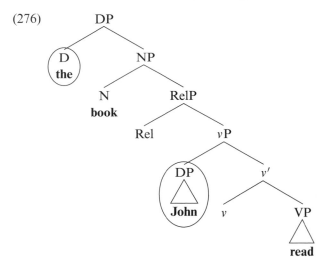

Since (following Krause) the CP phase of the relative clause is missing, the DP *John* and the D *the* are in the same Spell-Out domain in (276). We expect to see the same type of interaction between the higher D and the embedded DP that we have in the other cases reviewed in this section, and this seems to be the correct result; *John* in (276) must be made into a KP to prevent a Distinctness violation. Again, the general expectation of Distinctness is that we should not expect to see these phenomena only when DPs are in feature-checking or selectional relations with each other; any case of structural proximity between DPs should be handled in this way.

2.5.1.2 DP c-commanding DP Turning to the structure in (273b), in which one DP c-commands another, we found the same types of remedies for Distinctness violations; we can either add material to one of the DPs, making it into a KP, or remove the offending functional material, making one DP into an NP:

(277) a.

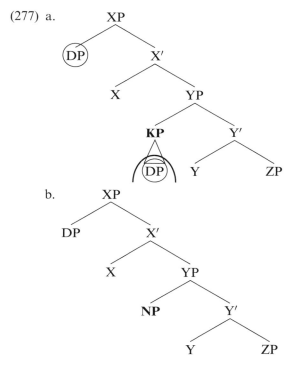

The first of these options is the one represented by differential case marking (section 2.2.1.2); the second might be involved in incorporation and pseudo-incorporation (Baker 1988, Massam 2001, and much other work).

Much work on morphological case crucially associates Case with the presence of multiple DPs; this is the idea of the Dependent Case approach (Massam 1985, Marantz 1991, and Harley 1995) and of Bittner and Hale's (1996) notion of Case competitors. On this type of approach, a DP is marked for Case just when there is another DP in the structural vicinity. The difference between nominative/accusative and ergative/absolutive languages has only to do with which of the two arguments in a transitive sentence is made into a KP; nominative/accusative languages Case-mark their objects, while ergative/absolutive languages Case-mark their subjects.[61]

The K(ase) head has played two roles in this chapter. In section 2.2.1.2, I claimed that K is a phase head in languages like Chaha, Hindi, Miskitu, and Spanish, preventing multiple DPs from being spelled out in the same Spell-Out domain, for example, when Object Shift brings a subject and an object too close together:

(278) Gɨyə **yə-fərəz** nəkʷəsənɨm (Chaha)
       dog **yə** horse bit
       'A dog bit a (specific) horse'

This type of language lends itself to an account in terms of Dependent Case; one of the arguments is a KP, and the other is apparently a DP.

Another class of languages appears to make most nominals into KPs. In sections 2.3.2 and 2.3.3, we saw evidence that in some languages, different values of case may be treated as different for purposes of linearization:

(279) a. [Sensei-o   hihansita] gakusei-ga   koko-ni oozei iru kedo,
           teacher-ACC criticized student-NOM here-DAT many be but
           dare-ga   dare-o   ka oboeteinai (Japanese)
           who-NOM who-ACC Q remember-NEG
           'There are lots of students here who criticized teachers, but I
           don't remember who who'
       b. *[Sensei-ga   suki na] gakusei-ga   koko-ni oozei iru kedo,
           teacher-NOM like   student-NOM here-DAT many be but
           dare-ga   dare-ga   ka oboeteinai
           who-NOM who-NOM Q remember-NEG
           'There are lots of students here who like teachers, but I don't
           remember who who'

The claim in section 2.3.2 was that (279a) is well formed because the two *wh*-phrases have different values for case, while in (279b) the *wh*-phrases have the same case and are therefore impossible to linearize.

There will have to also be instances of case morphology that is useless for rescuing Distinctness violations. English case morphology on pronouns is one instance of this:

(280) a. "It's raining," said **he** to **me**.
       b. *"It's raining," told **he me**.

In (280b), the two postverbal DPs have different values for case, but the result is still ruled out by Distinctness.[62]

We have seen reasons, then, to think that K can have (or lack) at least two properties: K can be a phase head, and K can come in different varieties that are treated as different for purposes of linearization. We might now reasonably wonder whether these properties are connected.

Preliminary investigation suggests that these properties must be dissociable at least in one direction; it must be possible for K to come in relevantly different varieties without being a phase head. The relevant data are from Dutch (Iatridou and Zeijlstra 2009). As we saw in section 2.3.2, Dutch is like German in allowing multiple sluicing with DP remnants:

(281) a. Ich habe jedem Freund ein Buch gegeben, aber ich weiß nicht
       I   have every friend a book given   but I know not
       mehr wem welches (German)
       more who which
       'I gave every friend a book, but I don't remember anymore
       who which'
    b. Iemand   heeft iets   gezien, maar ik weet   niet wie
       someone has   something seen   but   I know not who
       wat. (Dutch)
       what
       'Someone saw something, but I don't know who what'

Following the approach developed in section 2.3.2, we can take the well-formedness of (281b) as evidence that Dutch is capable of distinguishing between DPs, perhaps via different varieties of case. On this view, the multiple sluicing remnants in (281b) are linearized via a statement like ⟨[DP, NOM], [DP, ACC]⟩ (or perhaps ⟨[KP, NOM], [KP, ACC]⟩, if the Dutch nominals are instances of KP).

Dutch differs from German, however, in its treatment of nominal complements of nouns:

(282) a. die Zerstörung [der   Stadt] (German)
       the destruction the.GEN city
       'the destruction of the city'
    b. de verwoesting *(van) de stad (Dutch)
       the destruction   of   the city
       'the destruction of the city'

German nominals can be complements of nouns, but in Dutch, as in English, a nominal complement of a noun must be protected by insertion of a P.[63] By hypothesis, the German example in (282a) is well formed because German K (or at least genitive K) is a phase head:

(283)

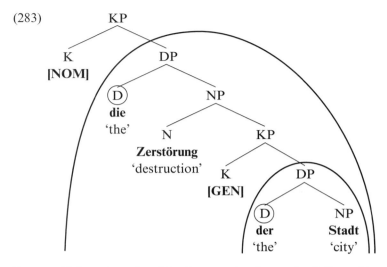

Because K is a phase head in German, it triggers Spell-Out of its complement. As a result, the functional heads in the extended projections of the nominals in (283) are linearized in separate phases, and no Distinctness violations are incurred.

We can give Dutch (282b) the same tree, as long as we say that Dutch K is not a phase head:

(284)

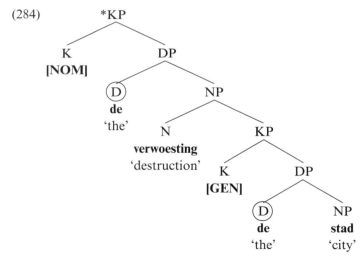

The tree in (284) violates Distinctness, not because of the two instances of K—by hypothesis, the multiple sluicing facts tell us that Dutch is capable

of distinguishing between different values of K—but because of the two
instances of D (along with whatever other functional heads we want to
posit in the extended projections of these nominals). By allowing Dutch
to distinguish between different values of K, but denying Dutch K the status of a phase head, we arrive at the result that Dutch KPs may be in the
same phase if they are in a c-command relation (as in the multiple sluicing example) but not if they are in a dominance relation (as in the nominal complement example). Thus, the Dutch facts seem to indicate that it
is possible for K to have different values that are relevant for linearization without being a phase head. I will have to leave for future research
the question of whether we also find languages in which K is a phase
head but does not come in values that can be treated as distinct for
linearization.

2.5.2 Case or Not?

An approach based on Distinctness has the virtue of making Case-driven
phenomena special instances of a general pattern; all of the examples in
(285) are instances of the same pattern, as far as this theory is concerned,
though only the first is handled by Case theory of the classic type.

(285) a. the destruction *(of) the city
      b. John was seen *(to) leave
      c. *a man who to dance with
      d. *I know everyone insulted someone, but I don't know who
          whom

A number of morphological alternations have become the center of con-
troversy over whether they involve Case or not. Tagalog and other languages of the "Philippine type" exhibit one example of this:

(286) a. Nagbigay ang magsasaka ng bulaklak sa
          NOM-gave ANG farmer   NG flower   SA
          kalabaw (Tagalog)
          water.buffalo
          'The farmer gave a flower to the water buffalo'
      b. Ibinigay ng magsasaka ang bulaklak sa kalabaw
          ACC-gave NG farmer   ANG flower   SA water.buffalo
          'A farmer gave the flower to the water buffalo'
      c. Binigyan ng magsasaka ng bulaklak ang kalabaw
          DAT-gave NG farmer   NG flower   ANG water.buffalo
          'A farmer gave the water buffalo a flower'

Tagalog sentences with multiple arguments typically exhibit alternations of this kind, in which morphology on the verb picks out one of the arguments, which is then marked with a special morpheme *ang*; DPs not marked with *ang* are marked with *sa* when dative, or *ng* otherwise.[64] The alternations have invited analysis in terms of Case, with the different verb forms being treated as different "voices" and *ang* being a marker of nominative (or absolutive) case; this was the account in Bloomfield 1917, for example. The difficulty with this type of analysis, as much research has shown, is that the alternations in (286) seem not to involve movement of the type classically associated with Case; if anything, the movements involved seem to be A′-movement (Schachter 1976, 1996; Guilfoyle, Hung, and Travis 1992; Richards 1993, 2000; Rackowski 2002; Aldridge 2004; and the references cited there). Much debate in the Austronesian literature has therefore focused on the question of whether these alternations are "really" associated with Case or not.

Distinctness allows us to short-circuit this type of debate, to some extent. The alternations in (286) may be like Case in that they are arrangements of functional heads designed to make structures linearizable. On the other hand, we have seen that a number of phenomena are driven by Distinctness in this way, including phenomena that have not classically been handled by Case theory. Both sides of the debate over alternations like those in (286) can therefore claim victory; the morphology shares properties with case morphology, but need not be identical to it syntactically.

A similar issue arises with Algonquian obviation morphology. In the Algonquian languages, sentences containing multiple third-person arguments[65] make a distinction among the arguments based on their roles in the discourse: one argument is required to be *proximate* (proximate arguments are, roughly, those which refer to the topic of discussion or "point of view" of the discourse) and all the others are marked as *obviative*:

(287) a. Washkeetôp nâw-âw mashq-ah (Wampanoag)
          man   sees-DIR bear-OBV
          'The man (topic) sees a bear'
      b. Washkeetôpâ-ah nâw-uq mashq
          man-OBV   sees-INV bear
          'A man sees the bear (topic)'

Possessed noun phrases, however, exhibit different behavior; the possessed noun must be marked obviative, and the possessor must be treated as proximate, regardless of the status of the referents of these nouns in the discourse:

(288) washkeetôp wu-hshum-ah (Wampanoag)
      man   3-daughter.in.law-OBV
      'the man's daughter-in-law'

Here we have another type of morphology that shows signs of being driven by Distinctness; in (287), the requirement apparently is that a functional head of a particular type be attached to one or the other of the two DPs, and in (288) the same morphology is used to separate a possessor from the possessee. Traditionally, the literature on Algonquian does not refer to this morphology as "Case," perhaps partly because of its discourse effects. Here, again, we have a type of morphology that has properties in common with case morphology as we traditionally understand it, but may not be completely equatable with Case. And again, Distinctness allows us to untangle the problem somewhat; like Case, obviation fixes potential Distinctness violations, involving nominals, but this role for obviation may not predict anything about its other syntactic properties.

2.5.3 Case Resistance

A version of Stowell's (1981) Case Resistance Principle may be made to follow from Distinctness. The Case Resistance Principle is meant to account for facts like (289):

(289) *They're talking about [that they need to leave]

The ill-formedness of (289) might be another case of Distinctness, if we assume, following Emonds (1985), that prepositions and complementizers have something in common. Consider this tree for part of (289):

(290)

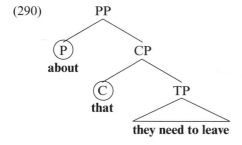

Recall that CP is a strong phase. When the phase above CP is spelled out, the nodes to be linearized will include P (*about*), C (*that*), and CP (*that they need to leave*). If P and C are effectively the same, then these nodes cannot be linearized; ⟨P, C⟩ is ruled out by Distinctness.

Here we arrive at a difficulty in the definition of "phase." Several earlier sections (particularly 2.1.3, 2.2.2.2, and 2.2.2.3) have relied on the
claim that PP is a phase. This was how we dealt with the contrast in
(291), for example:

(291) a. *the destruction [the city]
      b. the destruction [**of** the city]

Because the preposition *of* is a phase head, the reasoning went, (291b) can
avoid Distinctness by spelling out the DP *the city* in one Spell-Out operation, and then the higher DP *the destruction* in a later Spell-Out operation. If P and C are both phase heads in (290), however, we will lose the
result outlined above. Each phase head, by hypothesis, should trigger
Spell-Out of its complement, yielding the Spell-Out domains in (292):

(292)

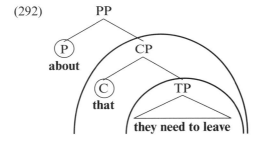

If these are the correct Spell-Out domains, then P and C should never be
linearized together. Some way of obviating the phasehood of P will have
to be found, if the ill-formedness of (290) is to follow from Distinctness.
   One way to distinguish between the P in (290), which should not be a
phase head, and the P in (291), which should, would be to link the behavior of P to that of *v*. In section 2.2.1.1, we crucially made use of Chomsky's (2000, 2001) claim that *v* is a phase head just when it is transitive. In
particular, drawing on Chomsky 2005, I proposed that *v* is like T in that
it is sometimes dominated by a phase head (which I called v_C) from which
it inherits the features necessary to Agree with a DP; thus, the phase head
v_C is Merged just when *v* is transitive, and needs to be able to Agree with
the object. This was the basis for the account of the contrast in (293):

(293) a. We saw John leave.
      b. *John was seen leave.

In (293a), the reasoning went, the matrix *v* is transitive, and is therefore
associated with a Spell-Out boundary, which it can cross via head move-

ment, thereby separating itself from the v of the lower clause. In (293b), by contrast, the matrix v is passive, and therefore the two instances of v are spelled out in the same Spell-Out domain, violating Distinctness.

Suppose we now extend this reasoning to P. Just when PP dominates a DP object, on this account, we must Merge a functional head P_C, from which P can inherit the features necessary to Agree with its object; moreover, P_C, like v_C and C, will be a phase head. This will make the desired distinction between the instance of P in (290) and the one in (291); the former will be dominated by the projection of a phase head, since its object is a DP, while the latter will not be.[66] We also correctly derive the fact (pointed out by an anonymous reviewer) that if a DP layer is inserted between P and CP, the result is well formed:

(294) a. *They're talking about [that they need to leave]
  b. They're talking about [**the fact** [that they need to leave]]

Because the P *about* in (294b) has a DP complement, it will be associated with a phase head, and hence will be insulated from the CP embedded in DP, making linearization possible.

Note that the Case Resistance Principle does not apply to interrogative clauses:

(295) They're talking about [what they should buy]

This is arguably related to the fact that such clauses behave as nominals when they are complements of nominals, in that they require *of* to be inserted:[67]

(296) the question *(**of**) [what they should buy]

Interestingly, Case Resistance effects seem to resurface in interrogatives when the *wh*-phrase is a PP. (297a) is somewhat stuffy-sounding in my dialect, but (297b) is ill-formed (and, I think, worse than (297c), suggesting that this is not simply a contrast of main clauses vs. embedded clauses):

(297) a. [With whom] should we discuss this?
  b. *They're talking about [with whom they should discuss this]
  c. They don't know [with whom they should discuss this]

The contrast in (297) is expected, again; in (297b), the P *with* and the PP *with whom* cannot be linearized:

(298)

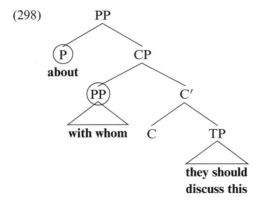

Here the objects to be linearized in the phase above CP include P (*about*) and PP (*with whom*). These cannot be linearized, since linearizing them would involve an ordering statement ⟨P, P⟩. The account is the same as that of the conditions on relative clauses developed in section 2.2.2.2:

(299) a.   a man [with whom to dance]
       b. *a man [who to dance with]

Example (299b) is ruled out because movement of *who* so close to the functional projections associated with the head noun creates a Distinctness violation; essentially, (299b) involves movement of a DP too close to another DP. The account of (297b) is the same; this example is ill-formed because a PP is brought too close to another PP.

Parts of Case theory, along with Case Resistance, then, can be made to follow from Distinctness. Classic Case theory still has a residue, which will be beyond the scope of Distinctness. We still need to understand, for example, what role Case plays, if any, in driving movement of DPs, in preventing such movement (via the Activity Condition, which prevents DPs from moving to Case positions once their Case features have been checked), and in conditioning the distribution of PRO. In all of these areas, proposals have been offered that eliminate the role of Case. Much work (see Marantz 1991 and Schütze 1997) has explored the idea that movement operations for DPs have nothing to do with Case per se, but are simply the result of general EPP requirements that can be filled by DPs in many instances. The Activity Condition has come under attack, for instance, in Nevins 2004, and the relationship between PRO and Case has also been questioned, in such works as Sigurðsson 1991, Landau 2004, and Bobaljik and Landau 2009. It may be possible, then, to dispense with any syntactic conditions that refer specifically to DPs and their licensing.

2.6   Conclusion

Because clauses often contain multiple DPs, Distinctness-related constraints on the relations between these DPs have often been observed,
and codified in the form of Case theory. One claim of this chapter has
been that much of Case theory is a special case of a more general theory.
In particular, facts like the one in (300a), which form part of the classical
motivation for Case theory, are special instances of a more general condition that is not limited to DPs (300b), nor is it crucially associated with
A-movement (300c):

(300) a.   the destruction *(of) the city
       b.   John was seen *(to) leave
       c.   *I know everyone insulted someone, but I don't know who
            whom

The more general principle of Distinctness, defended in this chapter, bans
Spell-Out domains containing more than one node of the same kind in an
asymmetric c-command relation. Such pairs of nodes will force the cre-
ation of linearization statements of the form $\langle \alpha, \alpha \rangle$, which are uninterpretable and cause the linearization process to crash.

   We have seen that defining the notion of "same kind" of node is not
trivial, and can be a point of crosslinguistic variation. In particular,
some languages (like English) invariably treat DPs as nodes of the same
kind, while others (like German and Japanese) assign different kinds to
DPs with different values for case or animacy. I have suggested that linearization statements make reference to feature bundles associated with
particular nodes, and that these feature bundles can vary in their richness
from language to language. The extent of this variance, and its consequences for other domains of grammar, is a topic for future work.

   We have seen that in many cases, a potential Distinctness violation
may be avoided in any of several ways, including removing offending
structure, adding additional structure to insert a phase boundary between
identical nodes, blocking movement operations that would create violations, or forcing movement operations that break up ill-formed structures. In sections 2.4.3 and 2.4.4.2.3 suggested that the grammar is
constrained by a general principle of avoiding Distinctness violations
wherever possible, and of eliminating them as rapidly as possible. Other
than in these brief discussions, I have had little to say about why particular languages choose particular options for coping with Distinctness; why
do some languages have construct state, for example, and others not? I
will have to leave this as a topic for future work as well.

3 Beyond Strength and Weakness

3.1 Introduction

The synonymous sentences in (1) illustrate a classic problem in syntax: some languages perform *wh*-movement overtly, while others do not:

(1) a. **What** did John buy __ ?
    b. John-wa **nani**-o   katta? (Japanese)
       John TOP what ACC bought

Throughout, I will refer to the option in (1b) as "covert movement," but in fact nothing I will say here will hinge crucially on the correct representation of this type of example. Much fruitful work has concentrated on how exactly to characterize the difference between (1a,b): should we understand Japanese as having roughly the same syntax for *wh*-questions as English, with a type of movement masked by some difference in the mapping of syntax onto phonology? or should we posit a different syntax for Japanese *wh* in situ, with a correspondingly different semantics for *wh*-words that allows them to be interpreted without movement?

While some progress has been made on these questions, comparatively little research has focused on the question of why languages differ in this way. The answer to this question that current theories standardly offer make the difference a parameter, something unpredictable about each language that must be learned by the child: in Minimalism, we speak of languages having "strong" or "weak" *wh*-features, or (more recently) of C having, or lacking, an EPP feature.

In this type of theory, the overt/covert distinction for *wh*-movement cannot be explained; it is simply stipulated, without following from anything else. Of course, this could turn out to be the right approach. In this chapter I will try, nevertheless, to find a deeper explanation; the goal here will be to predict whether a given language has *wh*-movement or *wh* in situ (or both). The proposal will be that the overt/covert distinction is indeed predictable from independently observable properties of languages; in particular, we can predict what a language will do with its *wh*-phrases from the position of its complementizer (particularly, the complementizer associated with *wh*-questions) and the nature of its mapping of syntactic structure onto prosody.

3.1.1   Japanese *wh*-Prosody

The proposal offered here is inspired by much recent work on the prosody of Japanese *wh*-questions (see Deguchi and Kitagawa 2002, Ishihara 2003, Sugahara 2003, Smith 2005, Hirotani 2005, and the references cited there). Pitch tracks for a Japanese statement, and a corresponding *wh*-question, are given in (2) (from Ishihara 2003, 53–54):

(2) a. Naoya-ga   nanika-o   nomiya-de nonda (Japanese)
       Naoya-NOM something-ACC bar-LOC   drank
       'Naoya drank something at the bar'

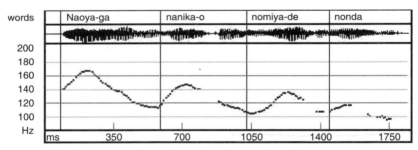

   b. Naoya-ga   nani-o   nomiya-de nonda no?
      Naoya-NOM what-ACC bar-LOC   drank Q
      'What did Naoya drink at the bar?'

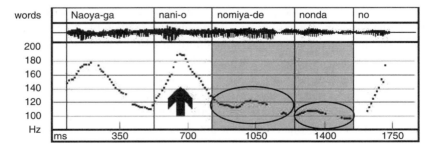

The pitch tracks in (2a) and (2b) differ in two ways. First, the *wh*-word direct object in (2b) has its pitch boosted; compare the lower pitch of the direct object in the statement in (2a). Second, there is a domain, starting with the *wh*-phrase and ending with the *wh*-complementizer (shaded in the pitch track in (2b), which is characterized by pitch compression: the peaks in this domain (circled) are lower than they would normally be.

Japanese *wh*-questions, then, involve a prosodic domain of some type which starts with the *wh*-phrase and ends with the complementizer. The proposal being defended here will be that all languages are attempting to do this; every language tries to create a prosodic structure for *wh*-questions in which the *wh*-phrase and the corresponding complementizer are separated by as few prosodic boundaries as possible. How languages achieve this varies from language to language, depending on where the complementizer is and on what the basic rules for prosody are.

Schematically, then, the proposal is this. Suppose we have an expression in which a *wh*-phrase and its corresponding complementizer are separated by prosodic boundaries, as in (3):

(3) C [$_\phi$] [$_\phi$][$_\phi$ wh]

There are two ways of satisfying the universal condition on *wh*-prosody being proposed here. One is to change the prosody of (3), creating a prosodic domain in which C and *wh* are not separated by prosodic boundaries:

(4) [C   wh]

As we will see, the option in (4) is available for some languages but not for others, and the distinction is predictable on the basis of independently observable properties of prosody. Another way of altering the structure in (3) to make it prosodically acceptable is to move the *wh*-phrase, so that it is closer to the C, in a position where no prosodic boundaries intervene between C and *wh*:

(5) [<u>wh</u> C [$_\phi$] [$_\phi$][$_\phi$ ~~wh~~]

The examples in (4) and (5) represent covert and overt movement, respectively.

At this point, two comments are in order. The first is that throughout this chapter I will make very unorthodox assumptions about the interaction between the syntax and the phonology; as we have just seen, the idea will be that the syntactic operation of *wh*-movement takes place just in case the prosody requires it. The approach therefore involves a straight-forward type of look-ahead.[1] We have the usual array of mechanisms for avoiding this problem: for example, we could allow the syntax to choose freely whether to move or not, or have the syntax create multiple copies in all cases, with the phonology acting as a filter ruling out certain derivations or choices of copy pronunciation. Alternatively, we could grant the syntax some way of knowing "in advance" certain facts about the phonological representation. Yet another option would be to simply keep our existing mechanisms for forcing overt and covert movement; the theory developed here would simply constrain the possible distributions of strong and weak features. One option which we can eliminate at the outset, I think, would be to make *wh*-movement a PF operation, taking it out of the syntax altogether. While this would solve the look-ahead problem, I think the evidence that *wh*-movement is syntactic is quite overwhelming. Choosing among the remaining options is not at all straightforward, however. For most of what follows I will simply ignore the issue, commenting on it only occasionally, as the theory is fleshed out.

The other comment has to do with the nature of the prosodic domain that we see in *wh*-questions in some languages (the example given above was from Japanese). The theory here will crucially constrain the distribution of prosodic boundaries, but it will have nothing to say about how these prosodic boundaries are realized as conditions on the rise and fall of pitch. In Japanese, as we have seen, the "*wh*-domain" is characterized by pitch compression, but there is no reason to expect this to be universal. In fact, we can see that it is not universal, without even leaving Japan. The pitch tracks discussed above were from Tokyo Japanese, but as Smith (2005) discusses, the facts are quite different in Fukuoka Japanese. Like Tokyo Japanese, Fukuoka Japanese has a special prosodic domain connecting *wh*-phrases with the complementizers where they take scope, but in Fukuoka Japanese the relevant prosodic domain is associated not with pitch compression but with a high tone. Pitch tracks for a non-*wh*-question and a *wh*-question are given in (6):

(6) a. Omae kyonen Kyooto itta   to   ya? (Fukuoka Japanese)
       you   last.year Kyoto went NLIZER COP
       'Did you go to Kyoto last year?'

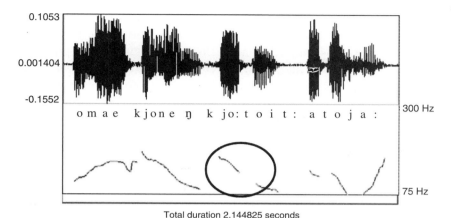

Total duration 2.144825 seconds

b. Dare-ga   Kyooto iku ka wakaran.
   who-NOM Kyoto   go Q know-NEG
   'I don't know who's going to Kyoto'

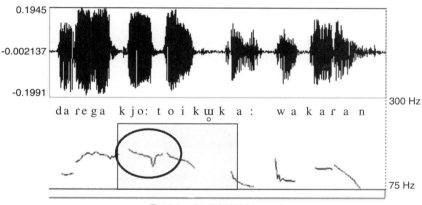

Total duration 2.218934 seconds

The pitch track for the word *Kyooto* 'Kyoto' is circled in both examples, and the *wh*-domain is shaded in (6b). As the pitch tracks show, *wh*-questions in Fukuoka Japanese involve not pitch compression but a high tone, starting at the beginning of the *wh*-domain and decaying somewhat toward the end. In the non-*wh*-question in (6a), *Kyooto* exhibits a dramatic fall, while in (6b), *Kyooto* is comparatively level and high.[2]

Thus, the claim being defended here is specifically related to the distribution of prosodic boundaries; as we will see, the claim will be that some but not all languages are capable of creating a "*wh*-domain" that captures the *wh*-phrase and the associated complementizer in a single domain, and these are the languages that can have *wh* in situ. What kind of effect these *wh*-domains have on pitch is not part of the theory: *wh*-domains might involve pitch compression, a high tone, or (in principle) no prosodic effects at all.[3] Similarly, we will see examples in which the complementizer that represents one edge of the *wh*-domain is phonologically null; again, the theory developed here is concerned with phonological representations rather than with phonetic effects.

3.2   Prosody and *wh*-Prosody

In order to understand why some but not all languages can derive the universally required prosody for *wh*-questions without moving the *wh*-phrase, we will first need to consider the conditions on prosody, and understand the ways in which mappings from syntax to prosody can differ from language to language.

Following much work on the syntax-phonology interface (see, for example, Selkirk 1984, Nespor and Vogel 1986, Truckenbrodt 1995, Wagner 2005, and much other work), I will assume that prosodic representations are constructed by mapping certain syntactic boundaries onto prosodic boundaries. For instance, we might have a language in which the Left edge of every DP is associated with a prosodic domain boundary. An SOV sentence in a language of this kind is schematically represented in (7a), and the corresponding prosodic structure is given in (7b):

(7)   a.   [TP ([DP D NP] [VP ([DP D NP] V]]
       b.   (D NP   )(D NP   V)

In (7), the Left edges of DPs (circled) are mapped onto prosodic boundaries, yielding a structure with two prosodic domains: the first consists entirely of the first DP, and the second of the second DP and the following verb.

(7b) shows the lowest level of phonological phrasing (what is sometimes called the *Minor Phrase* in the literature on prosody). Much work on prosody agrees that there are a number of domains of different sizes, which may be hierarchically organized, with Minor Phrases combining to form larger phrases (sometimes called *Major Phrases*), which are sometimes claimed to group together into Intonational Phrases, which in turn combine to form the Utterance. Selkirk's (1986, 1995) Strict Layer Hypothesis proposes that each of these levels of hierarchical phonological structure is always completely decomposable just into units of the next lower level:

(8)

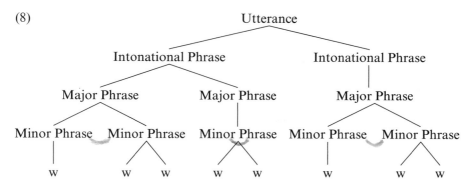

Very little of the diagram in (8) will be important for what follows. In fact, many aspects of the diagram in (8) have come under attack in recent work. For instance, Pak (2008) argues against the Strict Layer Hypothesis, discussing examples in which domains for phonological rules appear not to be in a reliable containment relation with each other. Ito and Mester (2007) offer arguments against dignifying the various levels of structure with distinct labels; in their model, phonological phrases are constructed recursively, but they claim that distinct labels for the different levels of structure are unnecessary. Whether (8) is the correct picture for prosodic structure is a matter that will not greatly concern us; the only level of importance to us will be the Minor Phrase, the phrase which immediately dominates the prosodic words.

In example (7), I made use of a rule that maps left edges of DPs onto prosodic boundaries. A great deal of research on prosody has been dedicated to discovering the correct form of this kind of rule; one perennial question, in particular, has to do with which syntactic categories these rules may refer to. I will not offer a principled answer to this question; I will address the question briefly again in section 3.3.3.2, where I will suggest that the maximal projections mapped onto prosodic domains might be the phases. For far more systematic investigation of the connection between phases and prosodic phrasing, see Dobashi 2004, Kahnemuyipour 2005, Ishihara 2007, Kratzer and Selkirk 2007, Pak 2008, and the references cited there.

I will follow Kubozono 2006 (and see Ishihara 2003 and Sugahara 2003 for much relevant evidence) in dealing with the facts of intonation in Japanese *wh*-questions by allowing Minor Phrases to be recursive, with multiple Minor Phrases being composed into a single, overarching Minor Phrase:[4]

(9)

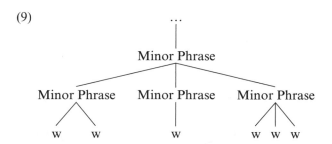

The algorithm for constructing these larger Minor Phrases will be the following:

(10) a. For one end of the larger Minor Phrase, use a Minor Phrase boundary that was introduced by a *wh*-phrase.

b. For the other end of the larger Minor Phrase, use any existing Minor Phrase boundary.[5]

Because the two Minor Phrases referred to in steps (9a) and (9b) need not be the same, the result of applying this algorithm will be a Minor Phrase which may (though it need not) consist of multiple smaller Minor Phrases. Suppose we consider again the phrasing given in (7), repeated here as (11), with the first DP changed to a *wh*-phrase:

(11) a. [$_{TP}$ [$_{DP}$ whP] [$_{VP}$ [$_{DP}$ D NP] V]]
b. (whP)(D NP V)

As we saw before, the rule "insert a Minor Phrase boundary at the Left edge of every DP" gives the Minor Phrasing in (7b), with the first Minor Phrase consisting of the first DP, and the second Minor Phrase consisting of the second DP and the following verb. Applying the process in (10) to the phrasing in (11b) would involve keeping the Minor Phrase boundary introduced by the Left edge of the *wh*-phrase, and the right boundary at the end of the utterance, yielding the larger Minor Phrase in (11'c):

(11') a. [$_{TP}$ [$_{DP}$ whP] [$_{VP}$ [$_{DP}$ D NP] V]]
b. (whP)(D NP V)
c. (whP D NP V)

The structure in (11') could be represented with the tree structure in (12):

(12)

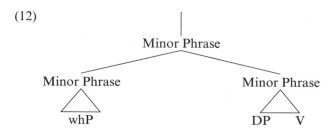

The prosodic effects associated with Japanese *wh*-questions, discussed in section 3.1.1, would then apply to the larger Minor Phrase, thus allowing the *wh*-phrase and the complementizer at the end of the clause to share a Minor Phrase. I will refer to these larger Minor Phrases that languages construct via the algorithm in (10) as "*wh*-domains."

The proposed condition on *wh*-prosody is then given in (13):[6]

(13) Given a *wh*-phrase α and a complementizer C where α takes scope, α and C must be separated by as few Minor Phrase boundaries as possible, for some level of Minor Phrasing.

How languages satisfy the condition in (13) will depend on how they distribute their Minor Phrase boundaries. Languages will be able to leave *wh* in situ just in case they have Minor Phrase boundaries placed in such a way as to be able to use the procedure in (10) to create larger Minor Phrases containing both the *wh*-phrase and the complementizer. Languages that cannot do this will try to improve the structure with movement operations, arranging for the *wh*-phrase and its complementizer to be closer together.

There will be two main points of crosslinguistic variation that will be relevant for us. One will be the position of the complementizer: complementizers may either precede or follow their complements. The other has to do with the placement of Minor Phrase boundaries, which can be either to the Left or to the Right of certain maximal projections. These two binary parameters leave us with four logical possibilities, which we will spend the rest of this section outlining schematically. The next section will go on to investigate in more depth the prosodic properties of one language of each type.

3.2.1 *Final* C, Minor Phrase Boundaries at *Left* Edges of XPs
The first possibility is the one sketched in the beginning of this section; a language with a final complementizer that places Minor Phrase

boundaries at Left edges of certain maximal projections. Such a language, as we saw, would associate the syntactic structure in (14a) with the Minor Phrasing in (14b). It would also be able to create a larger Minor Phrase containing the *wh*-phrase and the associated complementizer, by keeping the Minor Phrase boundary associated with the Left edge of the *wh*-phrase (circled), and skipping the immediately following one, as in (14c):

(14) a. [$_{DP}$] [**whP**][$_{DP}$] V C
     b. (   )(   )(   )
     c. (   )(   )

A language like this would therefore be able to leave *wh* in situ. We will see that Japanese is a language of this kind.

3.2.2 *Final* C, Minor Phrase Boundaries at *Right* Edges of XPs

A language like this would differ from Japanese in that it marks not Left edges, but Right edges of certain maximal projections with a Minor Phrase boundary:

(15) a. [$_{DP}$][**whP**] [$_{DP}$]   V C
     b. (   )(   )( )(   )(   )

As a result, it would be unable to create *wh*-domains of the type that Japanese uses. Recall that our algorithm for creating larger Minor Phrases says to take the boundary projected by the *wh*-phrase as one boundary for the larger Minor Phrase, and to use any Minor Phrase boundary on the other side as the other boundary. The Minor Phrase boundary associated with the *wh*-phrase (circled in (15) above) will always be to the right of the *wh*-phrase. Since the complementizer is final, a procedure that starts with keeping the boundary projected by the *wh*-phrase will be unable to improve the prosodic status of the *wh*-question; regardless of where the left boundary of the bigger Minor Phrase is, all the right boundaries intervening between the *wh*-phrase and the complementizer will be left intact. In the particular case of (15) (repeated as (15′)), our algorithm allows for the construction of the larger Minor Phrase in (15′c), which fails to improve the structure at all with respect to the condition on *wh*-prosody in (13):

(15′) a. [$_{DP}$][**whP**] [$_{DP}$]   V C
      b. (   )(   )( )(   )(   )
      c. (   )( )(   )(   )

A language like this, then, cannot create *wh*-domains; it will have to resort to *wh*-movement in order to improve the prosodic status of the structure. In particular, it will have to do everything possible to get the
wh-phrase further to the right, thus bringing it closer to the complementizer. We will see that this is the behavior of Basque.

3.2.3   *Initial* C, Minor Phrase Boundaries at *Left* Edges of XPs

The third case of interest is that of a language that is the mirror image
of Basque; in this language, both the complementizer and Minor Phrase
boundaries precede *wh*-phrases. Just as in Basque, this language will be
unable to create *wh*-domains, and for the same reason; our procedure for
creating *wh*-domains starts by keeping the Minor Phrase boundary projected by the *wh*-phrase, and in this case that Minor Phrase boundary is
between the *wh*-phrase and the complementizer, with the result that creating a larger Minor Phrase boundary will not improve the prosodic structure, as far as the conditions on *wh*-prosody are concerned. A sample
syntactic structure and its Minor Phrasing are given in (16a,b), with the
boundary projected by the *wh*-phrase circled; (16c) shows the results of
applying our procedure to create a larger Minor Phrase:

(16) a. C [$_{DP}$] [**whP**] [$_{DP}$]
     b. ( )(   )(   ) (   )
     c. ( )(   )(   )

As (16) shows, creating a larger Minor Phrase does not improve the structure; in particular, the *wh*-phrase is separated from C by just as many
Minor Phrase boundaries in (16c) as in (16b). This language therefore
cannot leave *wh* in situ; it will have to resort to *wh*-movement of the familiar type, moving *wh*-phrases to put them closer to C. We will see later
that Tagalog is a language of this type.

3.2.4   *Initial* C, Minor Phrase Boundaries at *Right* Edges of XPs

Finally, we arrive at a language that is the mirror image of Japanese, with
complementizers and Minor Phrase boundaries on opposite sides of *wh*-
phrases; in this case, the complementizer is initial, and Minor Phrase
boundaries follow their maximal projections. Just like Japanese, this type
of language will be able to leave *wh* in situ, by creating a larger Minor
Phrase containing both the *wh*-phrase and its associated complementizer;
in this case, the Minor Phrase will begin with the complementizer and end
with the *wh*-phrase:

(17) a. C [_{DP}] [**whP**] [_{DP}]
     b. (   ) (   )(   )
     c. (   )(   )

The language of this type to be discussed below will be Chicheŵa.

3.2.5 Predictions and a Hedge

In section 3.3 we will move on to outline the prosodic systems of Japanese, Basque, Tagalog, and Chicheŵa, and will see that they do indeed fall where I have put them in the typology above. Before we do that, however, we are already in a position to make at least two typological predictions.

One has to do with the relation between the position of the complementizer and the behavior of *wh*-phrases. Sections 3.2.1 and 3.2.2 discussed the two logically possible types of complementizer-final languages. What we have seen is that such languages will either leave *wh* in situ or will take whatever steps they can to bring *wh*-phrases further to the right. In other words, they will not have *wh*-movement of the traditional kind, moving *wh*-phrases to the left periphery of the clause.[7] This seems to be correct; verifiably complementizer-final languages seem to universally lack traditional leftward *wh*-movement:

(18) a. Taroo-wa **nani -o**   katta   no? (Japanese)
         Taroo-TOP what-ACC bought Q
         'What did Taroo buy?'
    b. Bkrashis-lags-kyis **gare** gzigs-gnang-pa-red pas? (Tibetan)
         Tashi-HON-ERG   what buy-do-PAST-AGR   Q
         'What did Tashi buy?'
    c. C'am^{w}ɨt **mɨr**   cək^{w}əracnɨm? (Chaha)
         C'am^{w}ɨt what cooked
         'What did C'am^{w}ɨt cook?'
    d. Qiaofong mai-le **sheme** (ne) (Chinese)
         Qiaofong buy-ASP what   Q
         'What did Qiaofong buy?'

Moreover, it does appear to be the complementizer that is the best predictor of *wh* behavior; *wh*-phrases remain in situ in complementizer-final languages, regardless of whether those languages are generally head-final (18a–c) or not (18d).

If complementizer-final languages with obligatory overt *wh*-movement are genuinely unattested, then I think we have a good argument against one of the options discussed above for dealing with the "look-ahead" problem inherent in the proposal developed here. Recall that one of the ideas discussed there was to simply retain our existing structure of strong and weak features regulating overt and covert *wh*-movement; on this view, the theory developed here would be a theory of the prosodic properties that well-formed questions must have in human language, but the syntactic computation itself would make no reference to these prosodic properties. The theory would simply be responsible for explaining why English, for example, must have a strong feature driving its *wh*-questions, while Japanese would have a weak one.

On this approach to the look-ahead problem, however, we would have no way of explaining the absence of complementizer-final languages with obligatory overt *wh*-movement to the left periphery of the clause. If Japanese did have a strong *wh*-feature, the questions created by *wh*-movement would be prosodically well-formed; Japanese is always capable of creating a *wh*-domain to connect a *wh*-phrase with an interrogative complementizer, over any distance, as long as the complementizer is to the right of the *wh*-phrase. *Wh*-movement does nothing to improve the prosodic structure of a Japanese question, but it would not harm the prosodic structure either.

What we appear to need, then, is a system that permits overt *wh*-movement just in case it improves the prosodic structure of the *wh*-question. In other words, we need the syntactic operation of *wh*-movement to take the prosodic consequences of its actions into account.

Another prediction of the theory developed here has to do with optionality of *wh*-movement. We have now seen two basic approaches to forming *wh*-questions; leaving the *wh*-phrase in situ, and moving it to put it closer to the complementizer. Some languages lack the prosodic means to leave *wh*-phrases in situ, and must do *wh*-movement. But for languages that have the option of leaving *wh* in situ, what we now expect is that, all other things being equal, *wh*-movement ought to also be an option, as long as the movement improves the prosodic structure of the question. Given that *wh*-movement and *wh* in situ both create prosodically acceptable structures, unless we add something to the theory to make one of these options preferable, both ought to be available, in principle. This seems to be the right result. It is most straightforwardly visible in complementizer-initial languages; what we expect is that even in those complementizer-initial languages that allow *wh* in situ (languages of the type discussed in section 3.2.4), *wh*-movement will also be an option. As far as I know, this is the case:

(19) a. qel   ʕaali štara   <u>ʔeeh</u>? (Egyptian Arabic)
        uncle Ali   bought what
        'What did Ali's uncle buy?'
     b. <u>ʔeeh</u> štara qel ʕaali?

(20) a. Tu as   vu   <u>**qui**</u>? (French)
        you have seen who
        'Who did you see?'
     b. <u>**Qui**</u> tu as vu?

Finally, let me end this section with a hedge. In the discussion of prosody above I have used the language of Selkirk's (1984) work and its many descendants, suggesting that languages may pick either right or left edges of certain maximal projections for mapping onto prosodic boundaries. I picked this particular approach since it is an influential one, and one which allows the predictions above to be outlined fairly straightforwardly. Even if it is at least partly incorrect, however (and it is far from being the only approach in the literature: see Nespor and Vogel 1986, Truckenbrodt 1995, 1999, Seidl 2001, Wagner 2005, Pak 2008, Selkirk 2009, and the references cited there for much discussion), I believe that the results discussed above should still stand. All that is necessary for this theory to work, as far as I can see, is an approach to prosody that allows prosodic phenomena to ultimately be sensitive to certain syntactic boundaries, and that allows the choice of relevant syntactic boundaries to vary crosslinguistically. The claim made above is essentially that *wh* in situ is a privilege reserved for languages that routinely associate prosodic phenomena with boundaries on a particular side of *wh*-phrases, and that have the complementizer on the opposite side of the clause. Any such theory will allow us to make generalizations about syntactic edges which routinely demarcate prosodic phenomena. The claim of this chapter is that the creation of *wh*-domains is governed by the language-specific principles that regulate prosodic phenomena more generally; once we know, for instance, that a given language standardly associates prosodic phenomena with the left edges of DPs, we can conclude that the language will also be able to demarcate *wh*-domains with left edges of DPs.

A variety of theories of the mapping between syntax and prosody are compatible with such a conclusion. I have described the relationship between prosodic phenomena and syntactic boundaries in terms of a mapping from syntactic structure onto prosodic structure, but this need not be the correct analysis. We could restate the account, for example, in terms of a theory of prosody that associated prosodic phenomena directly with syntactic structure, rather than with a prosodic structure derived from syntactic structure. Alternatively, we could imagine a theory with a more uniform mapping of syntactic structure onto prosodic structure than I have assumed here, which allows languages to make language-specific choices about which prosodic boundaries are relevant for the distribution of prosodic phenomena; such a theory might state, for instance, that languages universally map both edges of DPs onto prosodic boundaries, but that a given language may then elect to routinely use one or another of these edges for the organization of actual prosodic phenomena.

Similarly, the world of prosodic systems need not be as neatly symmetric as described above for this to be true; there may be more types of prosodic systems than I have outlined. If it turns out, for example, that some languages routinely mark *both* sides of maximal projections with prosodic boundaries, we would expect these languages to allow *wh* in situ. In such a language, it would always be possible to generate a *wh*-domain by taking one of the boundaries projected by the *wh*-phrase (crucially, the boundary projected on the opposite side of the *wh*-phrase from the complementizer) as one edge, and a domain next to the complementizer as the other edge.

In the case studies to follow, we will often encounter one particular type of complication; the Selkirk-style approach will work well for most types of phrases, but will fail for verbs or for verb phrases, which will turn out to have their own special phrasing. In some cases (e.g., Tagalog, Chicheŵa), phrase boundaries that we would expect to find in the neighborhood of the verb will be unaccountably absent; in others (e.g., Basque, Bangla), a special phrase boundary will be introduced to separate the verb from adjacent material with which it would otherwise be phrased. I will leave these mysteries as mysteries for now, hoping that at some point they will turn out to be instructive about the syntax of the clause, perhaps being related to verb movement in some way.[8] More research into the typology of prosodic systems should allow us to address these questions. For the time being, I will continue to work with the typology discussed above.

3.3 Case Studies (p. 189)

In what follows we will discuss the four types of languages outlined above in more depth.

3.3.1 Japanese: *Final* Complementizer, Minor Phrase Boundaries to the *Left* of Certain XPs

Japanese prosody is the subject of a rich and ongoing literature (see Poser 1984, Selkirk and Tateishi 1988, 1991, Ishihara 2003, Sugahara 2003, and the references cited there for much discussion). Here I will rely heavily on the work of Selkirk and Tateishi (1988, 1991).

Japanese Minor Phrase boundaries are signaled by a Low tone on the first mora of the phrase, a phenomenon known as Initial Lowering. Minor Phrasing is determined by several factors, of which the most syntactic in nature is a requirement that Left edges of certain maximal projections (including DPs) be mapped onto Minor Phrase boundaries.[9] This requirement can be illustrated with the ambiguous string of words in (21), which has either of the meanings in (21a) or (21b); in other words, *Oomiya-no* 'from Oomiya' can be taken to modify either *Inayama* '(Mr.) Inayama' or *yuujin* 'friend'. These readings correspond to the trees in (22a) and (22b) respectively:

(21) Oomiya-no   Inayama-no   yuujin (Japanese)
     Oomiya-GEN Inayama-GEN friend
     a. 'the friend of [Mr. Inayama from Oomiya]'
     b. 'Mr. Inayama's [friend from Oomiya]'

(22) a.

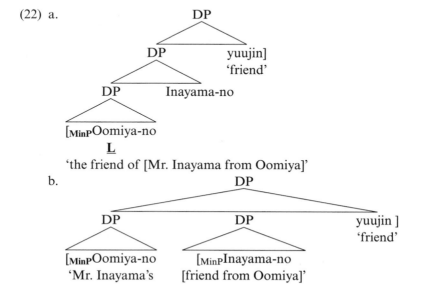

'the friend of [Mr. Inayama from Oomiya]'

The trees in (22) are annotated with their Minor Phrasing. The tree in (22b) must have a Minor Phrase break (and hence a Low boundary tone, realized on the first mora of the following word) between *Oomiya-no* and *Inayama-no*. The tree in (22a), by contrast, need not have such a prosodic break.

Selkirk and Tateishi (1988, 1991) propose that we capture this distinction between the trees in (22) by making reference to Left edges of maximal projections; in (22b), but not in (22a), a Left edge of a DP (namely, *Inayama-no*) intervenes between *Oomiya-no* and *Inayama-no*, and this Left edge is mapped onto a Minor Phrase boundary. Note that Left and not Right edges must be crucial; in both of the trees in (22), the Right edge of a DP (namely, *Oomiya-no*) appears in this position.

This mapping from maximal projection boundaries to prosodic boundaries is not the only condition on Minor Phrasing, but the other conditions seem less syntactic in nature. One of the conditions has to do with the distribution of lexical accent. Japanese has lexically accented words and lexically unaccented words, and one requirement on Minor Phrases is that they cannot contain more than one lexically accented word. Minor Phrase boundaries are therefore always inserted in such a way as to avoid this, whenever the situation might arise, regardless of whether the positioning of these boundaries marks edges of maximal projections. Minor Phrases are also subject to a length restriction; the specifics of this seem to vary from speaker to speaker and to be at least partly dependent on speech rate, but for some speakers, at least, Minor Phrases may not be more than three words long.

We have seen that Japanese is a language that routinely inserts Minor Phrase boundaries at Left edges of certain maximal projections; it also inserts Minor Phrase boundaries in some other positions, arguably positions that are not determined by the syntax. Japanese is also, famously, a head-final language, and in particular has final complementizers. In other words, Japanese places its Minor Phrase boundaries and its complementizers on opposite sides of potential *wh*-phrases. Given the theory outlined in section 3.2, this means that Japanese ought to be able to leave *wh* in situ; Japanese is a language that can create prosodic "*wh*-domains" by merging Minor Phrases to create superordinate Minor Phrases, which will begin with a Minor Phrase boundary projected by a *wh*-phrase and end at the complementizer where that *wh*-phrase takes scope. As we have seen, this is indeed the case.

3.3.2 Basque: *Final* Complementizer, Minor Phrase Boundaries to the *Right* of Certain XPs

In my discussion of Basque I will rely on Elordieta's (1997) analysis of the facts; see also Arregi 2002 and Gussenhoven 2004 for discussion.

Basque Minor Phrases are like their Japanese counterparts in that they begin with a Low tone, in this case realized on the first syllable of the first word of the Minor Phrase. Their distribution is quite different, however, as we will see. It will perhaps be easiest to discuss the Basque system by contrasting it with the Japanese one.

We saw in section 3.3.1 that Japanese always puts Minor Phrase boundaries at left edges of certain projections, including DPs. This is very far from being true in Basque; a Basque sentence may have a number of successive DPs with no Minor Phrase boundaries between them, regardless of the domination relations between them. All of the material before the verb in the examples in (23), for example, consists of a single Minor Phrase:

(23) a. [umiari]   [normalian] [urà]   [emoten dotzágu] (Basque)
         child-DAT normally   water give   AUX
         'Normally, we give water to the child'
     b. [Sure [erriko [alkatia]]] [Iruñara] [allaga
         our   town's mayor   Iruña-at arrived
         da] (Gussenhoven 2004)
         AUX
         'The mayor of our town has arrived in Iruña'

Both of these examples contain a DP in preverbal position that is c-commanded but not dominated by a preceding DP; in other words, these two DPs are linearly separated by both a Left edge of a DP (the second one) and the Right edge of a DP (the first one). Despite this, no Minor Phrase boundary intervenes between the two DPs. Neither a Left or a Right edge of a DP, then, is reliably mapped onto a Minor Phrase boundary in Basque. Instead, Minor Phrase boundaries appear in two types of positions.

Like Japanese, Basque distinguishes between lexically accented and lexically unaccented words. Unlike in Japanese, accented words in Basque must always be *followed* by a Minor Phrase boundary:

(24) a. [$_{MinP}$]   b. [$_{MinP}$] [$_{MinP}$]
     a'. lagunen   dirua   b'. lagúnen   dirua
         friend-GEN.SG. money   friend-GEN.PL. money
         'the friend's money'   'the friends' money'

The DPs in (24a′) and (24b′) have the minor phrasing in (24a) and (24b); because *lagúnen* 'friend-genitive plural' in (24b′) is lexically accented, it must be followed by a Minor Phrase boundary (with the result that the first syllable of *dirua* 'money' receives a low tone in (24b′)).

In Japanese, by contrast, lexically accented words need not be followed by Minor Phrase boundaries. Selkirk and Tateishi (1988) offer the phrasings in (25a^1) and (25a^2) as possible prosodic structures for the string of words in (25a), which begins with an accented word:

(25) a^1. $(_{\text{MinP}}$ $)(_{\text{MinP}}$ $)$
      a^2. $(_{\text{MinP}}$ $)(_{\text{MinP}}$ $)(_{\text{MinP}}$ $)$
      a′. [[[[Yamámori-no] yamagoya-no] uraniwa-no]

                                    umagoya-ni]]]]. . . . (Japanese)
         'In the barn in the backyard of a hut in Yamámori . . .'

Here Minor Phrasing is essentially free, dictated only by the need to make Minor Phrases sufficiently short. Thus, accented words in Japanese can be followed by a Minor Phrase break, but need not be.

The other position where Minor Phrase boundaries are required in Basque is immediately before the verb; that is, the first syllable of the verb must have the Low tone which signals the beginning of a new Minor Phrase. Here, again, Basque differs from Japanese, which allows the verb to phrase together with a preceding DP. The upshot of this is that for an SOV sentence consisting entirely of unaccented words, Basque and Japanese have the contrasting Minor Phrase structures in (26–27):

(26) a. $(_{\text{MinP}}$ $) (_{\text{MinP}}$ $)$ (Japanese)
      a′. *[subject] [object] verb*

(27) a. $(_{\text{MinP}}$ $)$ (Basque)
      a′. *[subject] [object] verb*

Basque, then, is unlike Japanese, in that it does not routinely place Minor Phrase boundaries at the Left edges of maximal projections. If anything, it seems to place its boundaries at the Right edges of certain projections. In particular, Basque puts Minor Phrase boundaries after lexically stressed words, and as a reviewer notes, we can understand the insertion of a boundary before the verb as marking the Right edge of the *v*P, if the verb has undergone head movement out of the *v*P to some high functional head.

It may well be that the prosodic behavior of lexically stressed words is irrelevant. Since lexical stress is crucially tied to particular lexical items, the prosodic boundaries associated with lexical stress may be placed postsyntactically, after lexical insertion is complete. In the introduction to this chapter I briefly raised the "look-ahead" problem inherent in the theory being developed here; how does the syntax make use of facts about the prosodic representation in order to decide whether to perform *wh*-movement overtly or covertly? I return to this question in section 3.6, but depending on the answer to the look-ahead question, we might hope that the syntax would be blind to postsyntactically inserted prosodic boundaries. For our purposes at the moment, however, the only important property of Basque prosody is that it is unlike Japanese prosody; Basque does not mark the Left edges of maximal projections.

Since Basque, like Japanese, is head-final, this means that Basque also differs from Japanese in that the complementizer is not routinely on the opposite side of *wh*-phrases from the Minor Phrase boundaries associated with those *wh*-phrases. Consequently, Basque ought to be unable to create the *wh*-domains that we find in Japanese. Recall that our algorithm for creating *wh*-domains begins by preserving the Minor Phrase boundary introduced by the *wh*-phrase; in Basque, if there is such a boundary, it will intervene between the *wh*-phrase and C, preventing the creation of a single Minor Phrase containing them both.

What should Basque do instead? What it in fact does is arrange for *wh*-phrases to be immediately preverbal:

(28) a. Mirenek **séin**   ikusi   rau? (Ondarroa Basque)
         Miren-ERG who-ABS see-PRF AUX.PR (Arregi 2002)
         'Who has Miren seen?'
     b. ***Séin**   Mirenek ikusi   rau?
         who-ABS Miren-ERG see-PRF AUX.PR

(29) a. Jon   **señek**   ikusi   rau?
         Jon-ABS who-ERG see-PRF AUX.PR
         'Who saw Jon?'
     b. ***Señek**   Jon   ikusi   rau?
         who-ERG Jon-ABS see-PRF AUX.PR

This requirement is expected under our theory. Consider, for example, the *wh*-phrase *séin* 'who-ABS' in (28). In (28b), where *séin* is not the immediately preverbal phrase, it is separated from the clause-final complementizer by two Minor Phrase boundaries: the one immediately following the lexically accented word *séin*, and the one right before the verb. In (28a), by contrast, only one Minor Phrase boundary (which is both immediately after *séin* and immediately before the verb) intervenes between the *wh*-phrase and the complementizer. The preference for the word order in (28a) over that in (28b) therefore follows.

On the other hand, we might expect to be able to improve the prosodic structure even more. If we take the facts about Basque prosody discussed above to indicate that Minor Phrase boundaries are established at Right edges of certain syntactic projections in this language, then we might expect Basque to move *wh*-phrases to postcomplementizer positions. The result would be a string like the one in (38):

(30) .... V C **wh**)

In (30), the *wh*-phrase, with its following Minor Phrase boundary, follows the complementizer, with the result that we ought to be able to construct a *wh*-domain linking the complementizer to the *wh*-phrase.

In fact, Basque is indeed capable of moving phrases to positions following the V-T-C complex (Elordieta 1997, 29):

(31) eweldi   onà   emon dábe mariñerúak (Basque)
     weather good give   AUX   fishermen-ERG
     'The fishermen have predicted good weather'

In (31), the DP *mariñuerúak* 'fishermen-ERG' follows the verb, and the attached complementizer. However, Elordieta (1997) reports that such postverbal material is always associated with radical pitch compression. In other words, postverbal material in Basque is already subject to conditions on prosody which might be incompatible with the conditions on *wh*-prosody being explored here.[10]

If we assume that movement to a postverbal position is ruled out on independent grounds, then the behavior of Basque is explained on the theory given here. The universal conditions on *wh*-prosody require Basque to minimize the number of Minor Phrase boundaries intervening between the *wh*-phrase and the associated complementizer. Unlike Japanese, Basque cannot achieve this simply by manipulating the prosody, for reasons that we have now derived from the basic rules for prosody in Basque; unlike Japanese, Basque does not routinely place Minor Phrase boundaries at Left edges of maximal projections, and therefore is not in a position to create a new Minor Phrase beginning with the *wh*-phrase and ending with the complementizer. Basque must therefore resort to movement, arranging for the *wh*-phrase to immediately precede the verbal complex in which the complementizer is located.

The Basque case is a potentially illuminating one for the nature of the prosodic requirements on *wh*-questions. As we saw in (28) and (29), Basque *wh*-phrases must be immediately preverbal. Arregi (2002) argues that this is accomplished by leftward scrambling of non-*wh*-phrases. Consider (29), repeated as (32):

(32) a. Jon   **señek**   ikusi   rau? (Ondarroa Basque)
         Jon-ABS who-ERG see-PRF AUX.PR
         'Who saw Jon?'
     b. ***Señek**   Jon   ikusi   rau?
         who-ERG Jon-ABS see-PRF AUX.PR

According to Arregi, the well-formed (32a) is created by scrambling the object *Jon* to the left of the subject *señek* 'who-ERG'. In other words, if Arregi is right, then the *wh*-phrase is put in the position required by the prosody, not by movement of the *wh*-phrase, but by movement of the non-*wh*-phrase. Basque does have scrambling, even in non-*wh*-questions, so the operation Arregi posits requires no new stipulations about Basque grammar.

In the account given here of Basque, this "altruistic" scrambling succeeds in improving the prosodic structure, though it does not make it perfect; there is still one Minor Phrase boundary between the *wh*-phrase and the corresponding complementizer. The condition on prosody will thus have to be stated as an economy condition, requiring the grammar to "do its best" to minimize the number of Minor Phrase boundaries between the *wh*-phrase and the complementizer; for reasons having to do with how Basque prosody works, Basque cannot fulfill this requirement perfectly, but the availability of scrambling allows the structure to be improved. The prosodic requirements do not empower the syntax to perform operations it cannot otherwise perform; it cannot move *wh*-phrases rightward into positions that are not there, or head-move the complementizer to a nonexistent initial position, for example.

Hypothetically, then, we might expect to find languages that are prosodically like Basque but differ from Basque in that they lack scrambling. Such a language would then be unable to improve the prosodic structure of its *wh*-questions at all; it would leave *wh*-phrases in situ (since moving them leftward would only make the situation worse), but would be unable to scramble non-*wh*-phrases out of the way to bring the *wh*-phrase closer to C.

Thus, we expect that *wh* in situ languages without scrambling (languages like Chinese, for example) might have the prosodic properties either of Japanese or of Basque; either is consistent with the approach developed here. Ultimately, of course, we hope to avoid stipulating that a given language either has or lacks scrambling, predicting this contrast from independently observable differences; for the time being, however, we will concentrate on developing a theory of this kind for *wh*-movement, leaving other types of movement for future work.

3.3.3   Tagalog: *Initial* Complementizer, Minor Phrase Boundaries to the *Left* of Certain XPs

Tagalog prosody has been the subject of comparatively little work (though see Schachter and Otanes 1972 as well as Kaufman 2005 for some discussion). What follows will be some of the results of a pilot study conducted at MIT.[11]

Various interesting properties of Tagalog will play no role in what follows. In particular, Tagalog verbs bear a type of morphology that has been the object of a great deal of study (see Richards 2000, Rackowski 2002, Aldridge 2004, and the references cited there for discussion; the facts are also briefly discussed in section 2.5.2), which refers to one of the arguments of the verb; it has sometimes been called "voice" morphology, I think misleadingly. Following Rackowski 2002, I represent this morphology with glosses like NOM (verbal agreement with the nominative argument), ACC, DAT, and so on. Nominals typically begin with case particles, glossed here with the relevant Tagalog morphemes ANG (nominal with which the verb agrees), SA (dative), NG (other).

Tagalog complementizers are initial, as we can see in the embedded clause of (33):

(33) Hindi ko   alam [kung   sumayaw   si   Maria] (Tagalog)
     not   NG.I know whether NOM-danced ANG Maria
     'I don't know whether Maria danced'

Also (and usefully for what is to follow) attributive adjectives in Tagalog may either precede or follow the noun they modify; a morpheme sometimes called the "linker," which I will gloss with LI, appears between them:

(34) a. Sumayaw   ang lolang   mayaman (Tagalog)
        NOM-danced ANG grandmother-LI rich
        'The rich grandmother danced'

b. Sumayaw   ang mayamang lola
   NOM-danced ANG rich-LI   grandmother

3.3.3.1 A Theory of Tagalog Prosody

We can begin our study of Tagalog intonation by considering the pitch tracks in (35) and (36). These two sentences both mean 'The weak servant drank the water', and differ only in the order of the adjective *mahina* 'weak' and the noun *alila* 'servant':

(35)

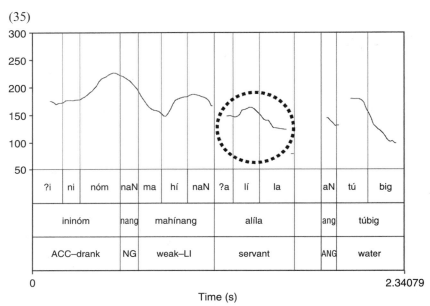

'The weak servant drank the water'

(36)

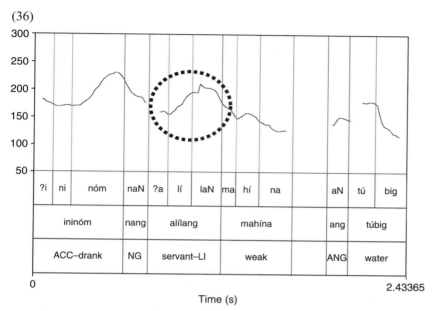

Time (s)

'The weak servant drank the water'

If we focus on the pitch track for the word *alila* 'servant' (circled in these examples), we can see that a given word need not always have the same intonation. In (35), *alila* has a peak on the second syllable, with the third syllable lower than the second. In (36), by contrast, the peak is on the third syllable. A similar effect can be seen on the adjective *mahina* 'weak', which has a clear fall in (36) that starts at the second syllable and continues through the third, while in (35) the pitch does not begin to fall until the end of the word.

Tagalog words, then, may either end in a fall or a rise; the final syllable may be either higher or lower than the preceding one. Considering the other words of these sentences, we can see that the verb *ininom* 'drank' consistently ends in a rise, while *tubig* 'water' consistently ends in a fall.

Since *tubig* 'water' is clearly phrase-final (since it is utterance-final), we might take the final-rise/final-fall distinction to distinguish between phrase-final and non-phrase-final words. On this theory, there is a phrase boundary just before *ang tubig* 'the water'. Phrases are marked in Tagalog by a phrase-final Low boundary tone, which causes phrase-final words to end in a fall. Thus, whatever word is just before *ang tubig* 'the water' in these examples is phrase-final, and must end in a fall. Non-phrase-final words, by contrast, end in a rise.

This assignment of phonological phrases also accounts for a pattern of downstep that we can see in these examples. The first three peaks in each example are successively downstepped, with each peak slightly lower than the preceding one. The last peak, the one on *tubig* 'water', is comparatively higher. Thus, phrase boundaries are apparently points of downstep reset in Tagalog. The preceding examples have the phrasing represented by the dark boxes in (37), with the verb and the subject phrased together, and a second phrase consisting just of the object:

(37)

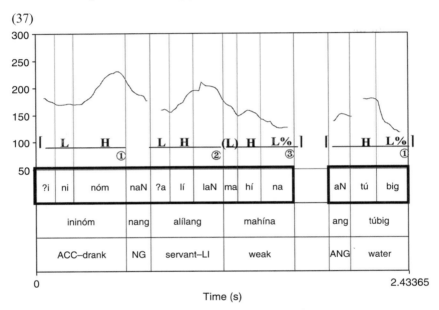

'The weak servant drank the water'

The first intonational phrase of (37) contains three content words, *ininom* 'drank', *alilang* 'servant-LI', and *mahina* 'weak'. The first two of these have their intonation peaks on the last syllable, but for the last one, the phrase-final Low boundary tone pushes the peak back to the second syllable. The second intonational phrase of (37) has only the content word *tubig* 'water'; being phrase-final, this word also has a penultimate pitch peak. This peak also exhibits downstep reset; each of the peaks in the first phrase is lower than the preceding one, but the peak of the second phrase is higher than the preceding one.

For these sentences, then, sentences with the word order VSO are phrased with the verb and the subject in one phrase and the object in a second phrase. There is much more to be said about the placement and nature of the pitch peaks in these words, but we can ignore this for now, concentrating instead on the question which concerns us; what is the algorithm for phonological phrasing? Two algorithms that would get the phrasing observed so far are given in (38):

(38) a. Place a phrase boundary at the right edge of every DP.
  b. Place a phrase boundary at the left edge of every DP, except for the one immediately after the verb.

Although the second of these algorithms is more complicated, I will now try to show that it is the correct one; in general, Tagalog places phonological phrase boundaries at left edges of certain maximal projections (including DPs), with the proviso that the verb must be phrased with immediately following material.

One argument for this has to do with the phrasing of postnominal possessors in Tagalog:

(39) Ininom [ng lolang mayaman [ni Maria]][ang
  ACC-drank NG grandma-LI rich NG Maria ANG
  tubig] (Tagalog)
  water
  'Maria's rich grandmother drank the water'

In a sentence like (39), the possessor *ni Maria* and the possessee *ng lolang mayaman* 'the rich grandmother' are separated by a left edge of a DP (namely, *ni Maria*), but not by a right edge of a DP. The algorithm in (38a), then, predicts that no phonological phrase boundary should precede *ni Maria*; the algorithm in (38b), on the other hand, predicts the presence of a phonological phrase boundary there. The second prediction is correct, as we can see by comparing the pitch tracks in (40) and (41):

(40)

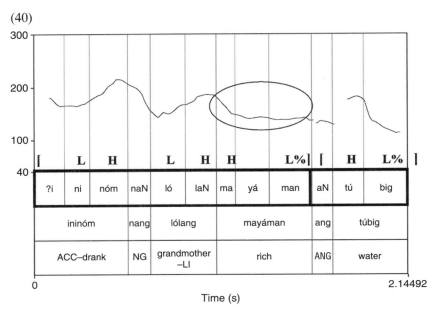

'The rich grandmother drank the water'

(41)

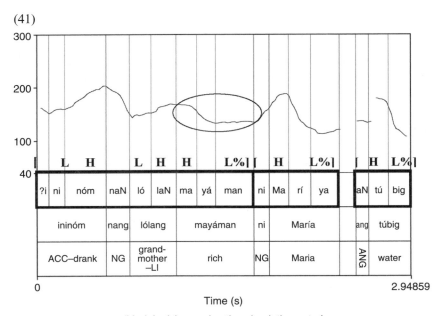

'Maria's rich grandmother drank the water'

(40) and (41) differ just in that (41) has a possessor, *ni Maria*, for the first DP. In both, the word just before the possessor (*mayaman* 'rich') has the pitch contour we expect in phrase-final position; its final syllable is low rather than high. In addition, the pitch peak on *ni Maria* in (41) is higher than the preceding one; in other words, the possessor exhibits reset of downstep, another test for phonological phrasing. Both tests argue for the presence of a phonological phrase boundary before the possessor. Again, this phrase boundary must be due to the presence of a Left edge of a DP (namely, the left edge of the possessor, *ni Maria*), since there are no Right edges of DPs that precede the possessor. We find the same result in longer sentences with multiply nested possessors:

(42)

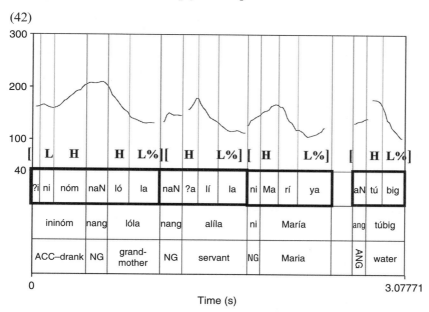

'Maria's servant's grandmother drank the water'

Here, *lola* 'grandmother', *ng alila* 'servant', and *ni Maria* all exhibit phrase-final prosody, characterized by their final Low tones (compare them with the prosody of the verb, the only word in this sentence which is not phrase-final; this one ends in a high tone, like all the non-phrase-final words we have seen).

The phrasing of possessors is one argument that Tagalog uses Left boundaries, rather than Right boundaries, to establish its phonological phrasing, and that the phrasing of the verb with the immediately following DP is the result of an overriding requirement that the verb not be in a phrase by itself. As we would expect on this theory, the verb's need to phrase with following material can be satisfied by a number of types of phrases. Adverbs, for example, can be phrased with the preceding verb:[12]

(43)

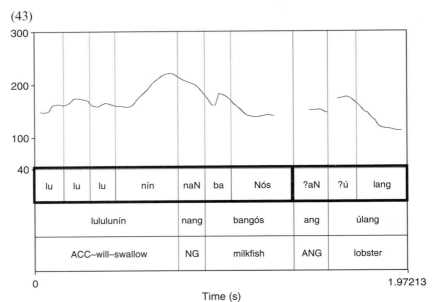

'The milkfish will swallow the lobster'

(44)

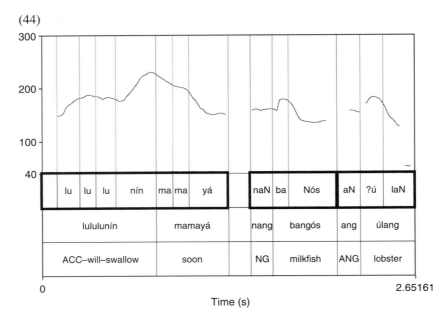

'The milkfish will swallow the lobster soon'

(44) differs from (43) just in the presence of a postverbal adverb, *mamaya* 'soon'. This adverb exhibits phrase-final prosody, ending in a Low tone; it is apparently phrased with the preceding verb, and followed by a phonological phrase break. This phrase break could be introduced by the Left edge of the following DP *ng bangos* 'the milkfish'; if phrase breaks are only introduced by Right edges of DPs, however, the facts in (44) are difficult to explain.

3.3.3.2 An Alternative Proposal A reviewer offers an alternative to the above account of Tagalog prosody. As he rightly points out, the theory given here relies on a brute-force characterization of the phrases that are mapped onto prosodic domains; I have simply declared here that DPs, and not other maximal projections, are associated with prosodic boundaries. The reviewer suggests that a more elegant principle for prosodic mapping in Tagalog would run as follows:

(45) In Tagalog, associate the Left edge of any *branching* maximal projection with a Minor Phrase boundary.

The principle in (45) uses the idea (see Nespor and Vogel (1986) and Uechi (1998) for discussion) that branching nodes—that is, nodes that dominate two separate pronounced constituents—have a special status for the syntax-prosody mapping. The reviewer notes that (45) yields the correct prosodic structure for examples like (41) above, repeated here as (46):

(46)

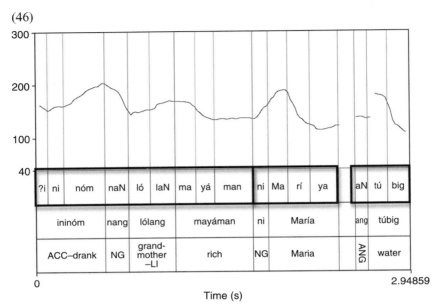

'Maria's rich grandmother drank the water'

The reviewer offers a tree essentially like the one in (47) for the sentence in (46). I have changed the reviewer's tree by adding branches for the morphemes *ng*, *ni*, and *ang*, which I have labeled K, and also by adding Asp(ect)P and *v*P:

(47)

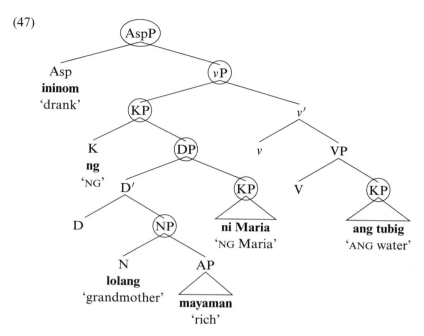

In this tree, the branching maximal projections have been circled. Note
that the NP *lolang mayaman* 'rich grandmother' has the N *lolang* and
the AP *mayaman* as its daughters, and hence is a branching node. Note
also that VP (for example) does not count as a branching node, in the
relevant sense, since only one of its branches is pronounced.

We can represent the tree in (47) via the labeled bracketing in (48); the
right brackets projected by branching nodes are the ones that are to be
mapped onto prosodic boundaries, according to the reviewer's theory,
and I have put them in boldface:

(48) [$_{Asp}$pininom [$_{vP}$[$_{KP}$ng [$_{DP}$[$_{NP}$lolang   mayaman] [$_{KP}$ ni
        ACC-drank   NG   grandmother-LI rich   NG
     Maria]]] [$_{VP}$[$_{KP}$ang tubig]]]]
     Maria   ANG water
     'Maria's rich grandmother drank the water'

Associating the right boundaries of branching nodes with Minor Phrase
boundaries gives us the following prosodic structure:

(49) (_{MinP}   )(_{MinP}

    [_{AspP}Ininom [_{vP}[_{KP}ng [_{DP}[_{NP}lolang   mayaman] [_{KP} ni

      ACC-drank   NG   grandmother-LI rich   NG

) (_{MinP}   )

    Maria]]] [_{vP}[_{KP} ang tubig]]]]

    Maria   ANG water

    'Maria's rich grandmother drank the water'

As the pitch track in (46) demonstrates, this is exactly the structure we want. The crucial property of the reviewer's theory is that it is capable of associating the edge of NP with a prosodic boundary; because the NP *lolang mayaman* 'rich grandmother' is branching, it is associated with a prosodic boundary to its right. As the reviewer points out, the alternative theory he suggests offers a general theory of which nodes are to be mapped onto prosodic boundaries; I have so far avoided offering such a theory, relying instead on arbitrary references to particular nodes. Moreover, the reviewer's theory, unlike mine, does not need to 'erase' any prosodic boundaries once they have been placed.

Unfortunately, although the reviewer's theory performs well for this particular example, it fails to capture all of the data discussed here. Consider the example in (42), repeated here as (50):

(50)

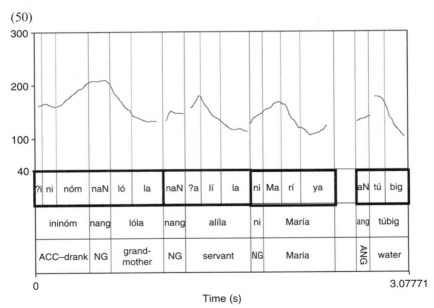

'Maria's servant's grandmother drank the water'

The sentence in (50) should have a tree like the one in (51); again, branching nodes are circled:

(51)

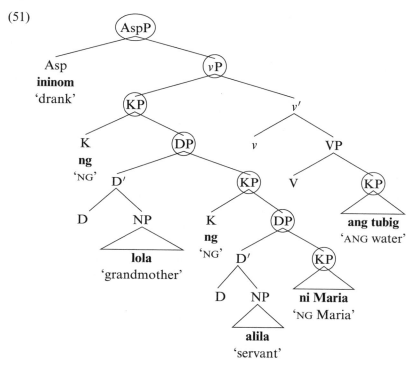

Again, we can represent the information in this tree with labeled brackets:

(52) [$_{AspP}$Ininom [$_{vP}$[$_{KP}$ng [$_{DP}$[$_{NP}$lola] [$_{KP}$ ng [$_{DP}$[$_{NP}$alila] [$_{KP}$ni
        drank   NG   grandma NG   servant NG
      Maria]]]]] [$_{VP}$[$_{KP}$ang tubig]]]]
      Maria   ANG water
      'Maria's grandmother's servant drank the water'

Associating Minor Phrase boundaries with the right edge of every branching maximal projection gives us the following result:

(53) ($_{MinP}$)
      [$_{AspP}$Ininom [$_{vP}$[$_{KP}$ng [$_{DP}$[$_{NP}$lola] [$_{KP}$ ng [$_{DP}$[$_{NP}$alila] [$_{KP}$ni Maria]]]]]
        drank   NG   grandma NG   servant NG Maria
      ($_{MinP}$)
      [$_{VP}$[$_{KP}$ang tubig]]]]
              ANG water
      'Maria's grandmother's servant drank the water'

The reviewer's algorithm divides this sentence into two Minor Phrases. But the pitch track in (50) shows that this is not the correct result; the subject should have Minor Phrase boundaries before each of the two possessors that it contains.

Thus, it appears that the reviewer's suggestion cannot cover all the Tagalog facts. On the other hand, I agree the reviewer that his proposal has one clear advantage over anything I have yet said. Rather than simply stipulating, as I have, that certain maximal projections and not others have their edges mapped onto Minor Phrase boundaries, he offers an algorithm for determining which maximal projections should be important for prosody; the algorithm associates Minor Phrase boundaries with the Right edges of all and only those maximal projections which immediately dominate multiple pronounced constituents (or, in the shorthand standardly used in the prosody literature, the maximal projections that "branch"). We have seen that this particular algorithm does not yield the right results, but it is clear that an algorithm is more desirable than a set of stipulations.

As the above examples show, the result that we want is to have Left edges of KPs (apart from the KP immediately following the verb) be associated with Minor Phrases. What special property of KPs could we appeal to to reach this result?

By the criteria developed in chapter 2, it seems clear that Tagalog is a language in which KP must be a phase. Arguments of nominals, for example, may be KPs:

(54) ang pagwawasak [ng lunsod] (Tagalog)
     ANG destruction   NG city
     'the destruction of the city'

If the reasoning in section 2.2.2.3 is correct, the KP *ng lunsod* 'NG city' must be a phase; Spell-Out of this KP protects the extended projection of the noun *lunsod* 'city' from being linearized together with the functional structure of the KP *ang pagwawasak* 'ANG destruction'.

We might capture the Tagalog facts, then, with the following generalization:

(55) Left edges of phases are mapped onto Minor Phrase boundaries
     (and the Minor Phrase boundary immediately following the verb is
     deleted).

This generalization still contains a stipulation about the verb, which I am not yet in a position to dispense with. But it is like the reviewer's suggestion in offering an algorithm, rather than just a stipulated list, for choosing maximal projections that are relevant for prosody. For more investigation of the idea that prosodic phrasing is directly linked to Multiple Spell-Out, see Dobashi 2004, Kahnemuyipour 2005, Ishihara 2007, Kratzer and Selkirk 2007, Pak 2008, and the references cited there. I return to the generalization in (55) in section 3.6.

Before concluding this section, I should also discuss a suggestion made by Sandra Chung (personal communication) for capturing the Tagalog data. She suggests that Tagalog could be described as a language which associates the Right edges of NPs with Minor Phrase boundaries. This proposal captures all of the Tagalog facts discussed here, without needing to postulate deletion of the phrase boundary immediately following the verb, as my proposal does. Because any KP will contain an NP, all KPs will come to have a Minor Phrase boundary at their right edge. If a KP contains a possessor, as in (56), the Right edge of the NP before the possessor will place a Minor Phrase boundary between the head noun and the possessor:

(56) $($MinP $)($MinP

$[$AspP$Ininom$ $[$vP$[$KPng $[$DP$[$NP$lolang$ $mayaman$ $]$ $[$KP$ ni$
 ACC-drank   NG   grandmother-LI rich   NG

     $)$ $($MinP $)$
 Maria $]]]$ $[$vP$[$KPang $tubig]]]]$
 Maria   ANG water
'Maria's rich grandmother drank the water'

Just like the reviewer's suggestion, then, Chung's proposal captures the fact in (56) by allowing the edge of an NP to be mapped onto a prosodic boundary, and thereby avoids having to posit deletion of the Minor Phrase boundary following the verb. Her proposal also shares with my original proposal one of the properties to which the reviewer objected; namely, that it simply stipulates that the edge of a particular category is associated with a prosodic boundary, rather than giving a general theory of how categories are mapped onto prosody. If the idea that phases are the categories mapped onto prosodic domains turns out to be a viable one, then we may be able to avoid this kind of stipulation.

Even if Chung's proposal turns out to be the best way to capture the Tagalog facts, we still predict that Tagalog wh-movement must be overt. Consider again the algorithm for building wh-domains, described in section 3.2:

(57) a. For one end of the larger Minor Phrase, use a Minor Phrase
         boundary which was introduced by a *wh*-phrase.
     b. For the other end of the larger Minor Phrase, use any existing
         Minor Phrase boundary.

The first half of this algorithm, (57a), refers crucially to "a Minor Phrase
boundary which was introduced by a *wh*-phrase." If Chung is right, then
Minor Phrase boundaries in Tagalog are in fact introduced not by KP
or DP, but by NP. In fact, however, it seems reasonable to think that
wh-phrases are not NPs, but KPs. It is certainly KPs which undergo *wh*-
movement:

(58) [$_{KP}$Sa aling   simbahan] mo   ibinigay ang pera? (Tagalog)
          SA which-LI church   NG.you OBL-gave ANG money
     'To which church did you give the money?'

In (58), it is clearly the KP that *wh*-moves; movement of any subpart of
this KP would be ungrammatical. If we were to use Chung's way of accounting for the facts of Tagalog prosody, then, we would arrive at the
result that Tagalog *wh*-phrases never project Minor Phrase boundaries at
all; *wh*-phrases are KPs, and in Chung's system, it is NPs that project Minor Phrase boundaries. Consequently, Tagalog should be unable to create
wh-domains, even under Chung's approach to Tagalog prosody.

In this section we have considered two alternatives to the account
of Tagalog prosody given in section 3.3.3.1. One, offered by a reviewer,
eventually failed to capture all of the Tagalog facts. The other, suggested
by Sandra Chung, captures all of the Tagalog facts, but shares with my
original proposal a need to stipulate particular nodes with which prosodic
phrase boundaries are to be associated. I have suggested here that my
original account can be made part of a more general theory of prosody,
in which the maximal projections whose edges are mapped onto prosodic
boundaries are specifically the phases. In what follows I will assume this
account.

3.3.3.3 Tagalog Prosody and Tagalog *wh*-Movement The arguments of the
previous two sections are meant to convince us that Tagalog prosodic
phrasing makes crucial reference to Left edges of certain maximal projections (notably DPs, or perhaps more accurately KPs). Tagalog, as we saw
earlier, is also a language in which the complementizer is initial. For the
theory developed here, then, Tagalog is effectively the mirror image of

Basque; both the complementizer and prosodic phrase boundaries will be on the same side of any given *wh*-phrase (the Left side, in this case; the Right side, in the Basque case). As in Basque, then, Tagalog will be unable to create prosodic *wh*-domains. Recall that our algorithm for creating these domains requires us to use one of the prosodic boundaries introduced by the *wh*-phrase as one of the edges of the new prosodic *wh*-domain, with the opposite boundary freely chosen from any of the existing boundaries. Since the prosodic boundary introduced by a Tagalog *wh*-phrase will be one on the Left edge of the *wh*-phrase, the opposite prosodic boundary will have to be to the right of the *wh*-phrase. Since the complementizer is initial, it will necessarily precede the *wh*-phrase, and the *wh*-domain will therefore not include the complementizer. Tagalog, then, is not like Japanese; it cannot meet the conditions on *wh*-prosody by leaving the *wh*-phrase in situ and manipulating the prosody. Tagalog ought to have obligatory *wh*-movement to the left periphery. And indeed it does:

(59) a.   **Kailan** umuwi   si   Juan? (Tagalog)
      when   NOM-went.home ANG Juan
      'When did Juan go home?'
    b.   *Umuwi si Juan **kailan**?

In fact, there is one case in which Tagalog does seem to allow *wh* in situ. Recall that Tagalog verbs must be phrased with immediately following material; the general requirement that Left edges of KPs be mapped onto phrasal boundaries is suspended in just this case. We might expect, then, that Tagalog would allow *wh* in situ just in case the *wh*-phrase in question was immediately postverbal; in such a case, no phrasal boundaries would intervene between the *wh*-phrase and the beginning of the sentence.

This is not as generally true as we might wish. Still, Tagalog does have one *wh*-word, *nino* 'who', which can appear just in the positions the theory leads us to expect:

(60) a.   Ninakaw **nino**   ang kotse mo? (Tagalog)
      ACC-stole NG.who ANG car   your
      'Who stole your car?'
    b.   *Ninakaw ang kotse mo **nino**?
    c.   *Sinabi   ng mga pulis [na ninakaw **nino**   ang kotse mo]?
      ACC-said NG PL   police LI ACC-stole NG.who ANG car   your
      'Who did the police say stole your car?'

As the data in (60) show, *nino* can appear in immediate postverbal position (as in (46a)), but cannot be separated from the verb by another KP (as in (60b)); moreover, the verb it follows must be the verb of the clause where it takes scope (cf. (60c)).[13]

This option of *wh* in situ is more constrained in Tagalog than we ought to expect, however. For one thing, *nino* actually sounds best in contexts in which it is not followed by any material at all; while (60a) above is acceptable, and clearly better than (60b), (61) is even better (particularly when the direct object is salient and hence easily dropped, as in a response to a declaration like 'someone has stolen my car'):

(61) Ninakaw nino? (Tagalog)
      ACC-stole NG.who
      'Who stole it?'

For another thing, *nino* appears to be the only *wh*-word to have this option:

(62) *Umuwi   kailan si   Juan? (Tagalog)
       NOM-went.home when   ANG Juan
      'When did Juan go home?'

It is unclear, then, how fully we should embrace this prediction of our theory with respect to Tagalog. More broadly, however, the theory seems to make the right prediction about Tagalog; it predicts that Tagalog should be a *wh*-movement language, as indeed it is.

(p. 186) 3.3.4   Chicheŵa: *Initial* C, Minor Phrase Boundaries to the *Right* of Certain XPs

The fourth and last case to be considered in this section is that of Chicheŵa; for work on the prosodic phrasing of Chicheŵa, see Bresnan and Kanerva 1989, Kanerva 1989, 1990, Truckenbrodt 1995, 1999, Seidl 2001, and the references cited there.

Kanerva (1989, 1990) discusses a number of tests for phonological phrasing in Chicheŵa. For example, he notes that the penultimate syllable of a phrase-final word undergoes lengthening (Kanerva 1990, 148):

(63) a. mtengo uuwu 'this price' (Chicheŵa)
       b. mteengo 'price'

Another rule, of Tone Retraction, retracts a phrase-final High tone to the penultimate syllable:

(64) a. mlendó uuyu 'this visitor' (Chicheŵa)
  b. mleéndo 'visitor'

Using tests like these, Truckenbrodt (1995, 1999) proposes that one of the factors determining phrasing in Chicheŵa is a mapping of Right edges of certain maximal projections onto prosodic domains. No phrase boundary intervenes, for example, between a noun and a following DP-internal maximal projection; the expressions in (65) consist of single phonological phrases (Truckenbrodt 1995, 76):

(65) a. ^{n}jingá yá mwáána (Chicheŵa)
   bicycle of child
  b. ^{n}jingá yábwiino
   bicycle good

Similarly, no phrase boundary intervenes between a preposition and a following DP; the expression in (66) is a single phonological phrase:

(66) ^{m}páká máawa (Chicheŵa)
  until tomorrow

Thus, left edges of PPs (as in (65a)), APs (as in (65b)), and DPs (as in (66)) are not mapped onto prosodic boundaries. On the other hand, subjects are separated from the VP by a phrase boundary, and coordinated DPs are also in separate phrases:

(67) a. [fíisi] [anadyá ḿkáa^{n}go] (Chicheŵa)
   hyena ate lion
  b. [miléeme] [^{n}di ^{n}júuchi]
   bats and bees

The data in (67) can be accounted for, Truckenbrodt points out, if we map right edges of DPs onto prosodic boundaries; the initial DP in both of these examples is separated from following material by this boundary.

Chicheŵa, then, is a language that maps Right edges of certain maximal projections onto prosodic boundaries. It is also a head-intial language, as the preceding examples reveal, with head-initial PP (as in (66)) and VP (as in (67a)). More importantly for our purposes, Chicheŵa complementizers precede the clauses they introduce (Bresnan and Kanerva 1989, 10):

(68) Zikugáníziridwá [**kútí** átsíbwéni ángá ndi afîti] (Chicheŵa)
  it.is.thought that uncle my be witch
  'It is thought that my uncle is a practitioner of witchcraft'

We have seen that Chicheŵa has initial complementizers and marks Right edges of certain maximal projections with prosodic boundaries. On the theory developed here, this makes Chicheŵa the mirror image of Japanese; a given *wh*-phrase will typically have the complementizer on one side (preceding it, in Chicheŵa; following it, in Japanese) and a phonological phrase boundary on the other (following it, in Chicheŵa; preceding it, in Japanese). Our algorithm for the creation of prosodic *wh*-domains thus allows Chicheŵa, like Japanese, to manipulate the prosody in ways that create a prosodically acceptable *wh*-question without movement. In other words, Chicheŵa ought to have the option of *wh* in situ. This is correct:

(69) anaményá **chiyáani** ndi   mwáálá (Chicheŵa)
      he.hit   what   with rock   (Downing 2005)
      'What did he hit with the rock?'

In fact, a variety of researchers (Kanerva 1989, 1990, Truckenbrodt 1995, 1999, and Downing 2005) have noted that Chicheŵa does manipulate the prosody of focus constructions and *wh*-questions in a way that is compatible with this theory. A difficulty for the account of Chicheŵa prosodic phrasing sketched here arises when we consider the phrasing of VPs; typically, VPs have no prosodic boundaries inside them. The VP in (70), for example, is a single prosodic phrase:

(70) (anaményá nyumbá ndí   mwáála) (Chicheŵa)
      he.hit   house   with rock
      'He hit the house with the rock'

If all Right edges of DPs are to be mapped onto prosodic boundaries, then we ought to expect to find a prosodic boundary after the DP *nyumbá* 'house', but this is not what we find. Truckenbrodt (1995, 1999) develops an Optimality-Theoretic account of the facts; in his approach, a constraint Wrap-XP demands that the VP be phrased as a unit, outranking the constraint that would normally place a prosodic boundary after the direct object (though see Seidl 2001 and McGinnis 2001 for criticisms of Truckenbrodt's approach).

Whatever account of the VP phrasing might turn out to be the right one, it is of interest that in *wh* in situ contexts, the expected phrasing can reassert itself:

(71) (anaményá **chiyáani**)(ndi   mwáálá) (Chicheŵa)
      he.hit   what   with rock   (Downing 2005)
      'What did he hit with the rock?'

Here the expected prosodic boundary after the direct object reappears. Recall that the algorithm for creation of prosodic *wh*-domains involves taking as one boundary for the *wh*-domain a prosodic boundary projected by the *wh*-phrase. In the case of (71), the relevant boundary is the one after the *wh*-phrase *chiyáani* 'what'. For whatever reason, this boundary is not ordinarily expressed in VP-medial position, but in *wh* in situ contexts it does appear, as we expect.

Here it may be useful to revisit the claim being made in this chapter about the prosodic representation of *wh* in situ. In section 3.2, I claimed, following Kubozono 2006, that the prosody of Japanese *wh* in situ should be represented as in (72):

(72) $[_{DP}$ $][_{DP}$ **wh** $][_{DP}$ $]$ C
   $(_{MinP}$ $)(_{MinP}$ $)(_{MinP}$ $)$
   $(_{MinP}$ $)$

Following much careful work on the prosody of Japanese *wh*-questions, I claimed that Japanese *wh* in situ triggers the creation of a '*wh*-domain', a larger prosodic phrase connecting the *wh*-phrase with the sentence-final complementizer, which is overlaid on the ordinary prosody of the sentence. In (72) above, this *wh*-domain is represented as the large Minor Phrase in the bottom line. As I showed in section 3.1, prosodic representations like the one in (72) receive different phonetic interpretations in different languages. In Tokyo Japanese, for instance, the prosodic structure in (72) is realized with pitch compression over the entire *wh*-domain, though the smaller Minor Phrase breaks represented on the second line of (72) can still be detected through careful acoustic work (see Ishihara 2003 and Sugahara 2003 for discussion of such work). In other dialects of Japanese, the *wh*-domain has different phonetic effects; the particular example we saw was Fukuoka Japanese, in which the *wh*-domain is associated with a high tone. As I pointed out in section 3.1, the claim of this chapter is about phonological representations like the one in (72), and not about their phonetic realizations, which can vary from language to language. In fact, as far as this theory is concerned, there could be languages in which the *wh*-domain has no phonetic effects at all.

In this section I have claimed that Chicheŵa *wh* in situ should be associated with a prosodic representation very similar to the Japanese one in (72):

(73) C $[_{DP}$ $][_{DP}$ **wh** $][_{DP}$ $]$
   $(_{MinP}$ $)(_{MinP}$ $)(_{MinP}$ $)$
   $(_{MinP}$ $)$

The phonetic facts from Chicheŵa discussed above amount to evidence that the prosodic boundary following the *wh*-phrase is always retained, even in contexts, like the middle of a VP, in which Chicheŵa generally deletes prosodic boundaries (or, at any rate, does not express such boundaries phonetically). The proposal was that the prosodic boundary after the *wh*-phrase is retained, even in such contexts, in order to make the creation of a *wh*-domain possible.

A reviewer notes (and see also Wagner 2005 and Dobashi 2006 for discussion) that *wh*-domains like the one in Chicheŵa, which extend to the left of the phrase that triggers them, appear never to be phonetically realized with pitch compression, in the way that the Tokyo Japanese *wh*-domain is. If true, this is a very interesting asymmetry, though not one I have anything to say about. The claims defended here would certainly be bolstered if a *wh*-domain of the Chicheŵa type, in some language that exhibits it, could be shown to have some phonetic effect on all the material inside it.

3.4   Interlude: More Wrap

In the preceding sections, I developed a theory that predicts, for any given language, whether that language will be able to create prosodic *wh*-domains, and hence be able to leave *wh*-phrases in situ. I claimed that we can predict this from the position of the complementizer, together with the general alignment of prosodic boundaries. Languages in which the complementizer is on one side of the clause, and prosodic boundaries are associated with the opposite side of certain maximal projections, I claimed, are the languages that can create *wh*-domains. Thus, we find *wh*-domains in languages like Japanese, in which the complementizer is on the Right edge of the clause and prosodic boundaries are associated with the Left edges of DPs, and also in languages like Chicheŵa, in which the complementizer is on the Left edge of the clause and prosodic boundaries are associated with the Right edges of DPs.

A reviewer suggests that this proposal is doomed to eventual failure. A more comprehensive investigation of prosodic systems, he claims, will eventually find fatal counterexamples. He proposes a third parameter distinguishing between languages, having to do with the constraint Wrap that appeared in the discussion of Chicheŵa, in section 3.3.4. (p. 182)

We saw in that section that the Chicheŵa VP typically contains no prosodic boundaries (though, as we also saw, these prosodic boundaries can reappear if required to do so by focus). As we noted, this fact has been captured by Truckenbrodt (1995, 1999) by means of a constraint Wrap, which demands that the VP be phrased as a single prosodic domain. In a language like Chichewa, on Truckenbrodt's account, Wrap outranks the constraints that would ordinarily align the edges of the VP-internal DPs with prosodic boundaries, causing those boundaries to disappear.

The reviewer suggests that the typology developed above should be supplemented with another parameter; some languages have the Wrap constraint ranked high in an Optimality-Theoretic hierarchy of constraints, and others do not. Only languages with high-ranking Wrap, the reviewer suggests, may construct *wh*-domains. Thus, the reviewer's idea would restrict further the class of languages in which *wh* in situ is possible. In order to have *wh* in situ, if the reviewer is correct, languages must not only have the properties described in the previous sections, but must also have high-ranking Wrap.

The reviewer may well turn out to be correct in his claim that the proposal developed in the preceding sections is empirically inadequate. The theory will clearly have to be tested against a wider range of languages. In the remainder of this section, I will try to show why we should not immediately adopt the reviewer's suggestion, though of course we may discover the need for something like it in future work.

Truckenbrodt (1995, 1999) and Seidl (2001) discuss a number of Bantu languages, which vary in their ranking for Truckenbrodt's Wrap constraint; that is, some of these languages are like Chichewa in that the VP is typically phrased as a single unit, and others are not. They offer, for example, the following phrasings for VPs containing two DPs (see Truckenbrodt 1995, 1999, and Seidl 2001 for arguments for these structures):

(74) a. (V DP)(DP) (Swahili)
          (Seidl 2001)
     b. ((V DP) DP) (Kimatuumbi)
          (Truckenbrodt 1999)
     c. (V DP DP) (Kikuyu)
          (Seidl 2001)
     d. (V DP DP) (Chichewa)
          (Truckenbrodt 1999, Seidl 2001)

All of these languages assign prosodic boundaries to Right edges of maximal projections. On Truckenbrodt's theory, they differ in the ranking of Wrap. In Swahili, Wrap is ranked quite low, so that it is outranked by the constraint which aligns maximal projections with prosodic boundaries, and we therefore find a prosodic boundary between the two VP-internal objects. The other three languages listed above have Wrap highly ranked; in Kimatuumbi, on Truckenbrodt's theory, Wrap is so highly ranked that it outranks another constraint which militates against recursive structures, and in Kikuyu and Chicheŵa Wrap outranks the constraint that rewards alignment of maximal projections with prosodic boundaries.

Although these languages differ in how they rank Wrap, they are identical in one respect: they all allow *wh* in situ:

(75) a. A-na-taka   u-tumi-e   dawa
       1s-PRES-want 2ss-take-SUBJ medicine
       gani? (Swahili)
       what (Hinnebusch and Mirza 1979)
       'What kind of medicine does he want you to take?'
    b. A-tel-įke   námanį? (Kimatuumbi)
       1s-cook-ASP what (Odden 1996, 62)
       'What did he cook?'
    c. Abdul a-ra-nyu-ir-ɛ   kee? (Kikuyu)
       Abdul 1s-TNS-drink-ASP-FV what (Schwartz 2007, 140)
       'What did Abdul drink?'
    d. A-na-mény-á chiyáani ndi   mwáálá? (Chicheŵa)
       1s-TNS-hit-FV what   with rock (Downing 2005)
       'What did he hit with the rock?'

It appears that the Bantu languages quite generally allow *wh* in situ, regardless of the ranking of Wrap. Thus, the reviewer's proposal, that a language must have high-ranked Wrap in order to construct *wh*-domains (and hence, license *wh* in situ), seems unsupported, at least in its simplest form.

Of course, one could imagine positing multiple versions of Wrap, which would be responsible for different parts of the clause, or for different levels of prosodic structure. In fact, there could be a version of Wrap that was responsible solely for the creation of *wh*-domains. Such a version of Wrap would amount to a stipulation that certain languages allow *wh* in situ and others do not.

As I mentioned at the outset of this chapter, this type of stipulation has a long tradition in the theory, and it may indeed turn out to be the best we can do. For the time being, however, I will continue to pursue the hope that we can construct a predictive theory of the distribution of strategies for *wh*-question formation.

3.5 Possible Further Directions

⟨p. 157⟩

In section 3.3, I discussed one language from each of the four possible types outlined by this theory. The languages are given in the table in (76):

(76)

	C to right of TP	C to left of TP
Prosodic boundaries on right of XPs	**Basque**	**Chicheŵa**
Prosodic boundaries on left of XPs	**Japanese**	**Tagalog**

Wh-questions in these languages are constrained by a requirement that the *wh*-phrase be separated from the complementizer where it takes scope by as few Minor Phrase boundaries as possible, for some level of Minor Phrasing. An algorithm for the creation of new Minor Phrases allows some languages to satisfy this condition without movement; the algorithm is given again in (77) (repeated from (10)):

(77) a. For one end of the larger Minor Phrase, use a Minor Phrase boundary which was introduced by a *wh*-phrase.
    b. For the other end of the larger Minor Phrase, use any existing Minor Phrase boundary.

Whether the algorithm in (77) can improve the prosody of the structure depends on whether the Minor Phrase boundaries introduced by *wh*-phrases intervene between *wh*-phrases and the associated complementizer or not. In the shaded cells in the tree in (76), they do not; the complementizer and the phonological phrase boundaries associated with maximal projections are on opposite sides of the *wh*-phrase. These are therefore languages that allow *wh* in situ, since their typical rules for creation of prosody allow the creation of acceptable prosodic structures for *wh*-questions. The languages in the unshaded cells, by contrast, must resort to movement to improve the structure; each must do whatever it can to move the *wh*-phrase and the associated complementizer closer together. In the case of Basque and languages like it, the result of this will be that the *wh*-phrase must be as far right as possible (immediately preverbal, in the particular case of Basque), while in the more familiar case of Tagalog, *wh*-movement will be to the left periphery of the clause.

In what follows we will consider how this general approach might be extended to some other cases.

3.5.1 Spanish

Uribe-Etxebarria (2002) and Reglero (2005) note an interesting condition on *wh* in situ in some dialects of Spanish. For some speakers, *wh* in situ is apparently acceptable in examples like (78):

(78) a. Juan compró qué? (Spanish)
            Juan bought what
       b. Tú le   diste la guitarra a quién?
            you CL gave the guitar   to who

Crucially, these are examples in which the *wh*-phrases are utterance-final. The *wh*-phrase may also be utterance-medial, for these speakers, but must be followed by a 'pause':

(79) Tú le   diste a quién *(#) la guitarra? (Spanish)
       you CL gave to whom   the guitar

The example in (79) cannot be given the ordinary non-*wh* intonation, in which the indirect object is immediately followed by the direct object; the *wh*-phrase must be followed by an intonation break.

Here I am hampered by my ignorance of Spanish intonation. But one possibility is that Spanish is essentially like Chicheŵa: a complementizer-initial language in which *wh* in situ is possible, just when a prosodic break appears just after the *wh*-phrase. The phonetic implementation of the prosodic break in Spanish is not the same as the one in Chicheŵa, but this is a situation we have already seen in section 3.1.1, when we compared *144* the *wh*-intonation of Tokyo Japanese to that of Fukuoka Japanese. The theory under development here is concerned only with prosodic structure, not with phonetic implementation.

3.5.2 Bangla

Bayer (1996), Simpson and Bhattacharya (2003) discuss conditions on the position of *wh*-phrases in Bangla (also called Bengali). Descriptively speaking, it appears that *wh*-phrases must linearly precede the complementizer where they take scope. Before we begin discussing Bangla, however, it will be useful to return briefly to Japanese.

Although Japanese is quite straightforwardly head-final, it does have a form of extraposition that can move phrases to postverbal position, as we saw in section 2.3.2 (and see Endo 1996 and Murayama 1998 for further discussion):

(80) John-wa katta yo, ano hon-o (Japanese)
John-TOP bought ASSERTION that book-ACC
'John bought (it), that book'

Clauses, both declarative and interrogative, can also be extraposed in this
way (Takako Iseda, personal communication):

(81) a. John-wa shinjiteiru yo, [Mary-ga dokushin da
John-TOP believes ASSERTION Mary-NOM single is
tte] (Japanese)
that
'John believes that Mary is single'
b. Keesatsu-wa shirabeteiru yo, [dare-ga okane-o
police-TOP investigating ASSERTION who-NOM money-ACC
nusunda ka]
stole Q
'The police are investigating who stole the money'

And a matrix clause which is itself a question may exhibit rightward
extraposition of an embedded clause:

(82) John-wa shinjiteiru no, [Mary-ga dokushin da tte]? (Japanese)
John-TOP believes Q Mary-NOM single is that
'Does John believe that Mary is single?'

However, a *wh*-phrase in an extraposed clause may not take matrix scope:

(83) *John-wa shinjiteiru no, [dare-ga dokushin da tte]? (Japanese)
John-TOP believes Q who-NOM single is that
'Who does John believe [__ is single]?'

Without extraposition, of course, such a sentence is well formed:

(84) John-wa [dare-ga dokushin da tte] shinjiteiru no? (Japanese)
John-TOP who-NOM single is that believes Q
'Who does John believe [__ is single]?'

In Japanese, then, just as our theory predicts, *wh*-phrases are required to
linearly precede the complementizer at which they take scope. Rightward
extraposition moves phrases to the right of the matrix complementizer,
with the consequence that rightward extraposition of *wh*-phrases, or of
any phrase containing *wh*-phrases, makes matrix scope for those *wh*-
phrases impossible.

Next we can turn to Bangla. Bangla is generally head-final (see Bayer 1996, 252–254, for arguments and discussion; Bangla has postpositions, and verbs that follow most of their complements except in marked orders). However, complement clauses may either precede or follow their selecting verb (Bayer 1996, 254):

(85) a. chele-Ta   jane   na [baba aSbe] (Bangla)
         boy-CLASSIFIER know.3 not father will.come
         'The boy doesn't know that (his) father will come'
     b. chele-Ta [baba aSbe] jane na

One difference between Bangla and Japanese is that the matrix complementizer in Bangla is null, which means that we cannot directly determiner the relative order of the matrix complementizer and the extraposed clause. In Japanese, as we saw, the extraposed clause is verifiably to the right of the matrix complementizer ((82), repeated as (86)):

(86) John-wa shinjiteiru no, [Mary-ga   dokushin da tte]? (Japanese)
      John-TOP believes   Q   Mary-NOM single   is that
      'Does John believe that Mary is single?'

As we will see, the Bangla facts will follow if we assume that Bangla and Japanese are the same in this respect.

In preverbal complement clauses, Bangla wh-phrases may take either embedded or matrix scope (Bayer 1996, 272):

(87) ora [**ke** aSbe]   Suneche (Bangla)
      they who will.come heard
      'Who have they heard will come?' (*matrix*)
      'They have heard who will come' (*embedded*)

In postverbal complement clauses, by contrast, Bangla wh-phrases must take embedded scope (Bayer 1996, 273):

(88) ora Suneche [**ke**   aSbe] (Bangla)
      they heard   who will.come
      'They have heard who will come' (*embedded* only)

For some Bangla speakers, at least, wh-phrases from postverbal clauses may take matrix scope by being moved into the matrix clause (Simpson and Bhattacharya 2003, 133[14]):

(89) jOn **ke**   bollo [__ cole gEche]? (Bangla)
      John who said   left gone
      'Who did John say left?'

The generalization, then, appears to be that a *wh*-word in Bangla may take scope in a clause just if it linearly precedes the verb (and, this theory leads us to hope, the postverbal complementizer) of that clause.[15] *Wh*-phrases from embedded clauses may take matrix scope just if the embedded clause precedes the matrix verb (as in (87)) or if the *wh*-phrase itself is moved to a preverbal position in the matrix clause (as in (89)). Otherwise, only embedded scope is possible (as in (88)).

This pattern can be made to fit into the theory developed here, on two additional assumptions. One is about the nature of Bangla prosody, which will need to be relevantly like that of Japanese—that is, it will have to insert prosodic phrase boundaries at Left edges of certain maximal projections. The other is about the position of the complementizer to which *wh*-phrases are to be related. This will have to immediately follow the verb of its clause (again, just as in Japanese).

We will shortly see some evidence for the first of these assumptions, but first let us consider how the assumptions help us to derive the facts. Consider first the possible readings of an example like (87), repeated as (90) (with the phonologically null complementizers added, just where they would be in the corresponding Japanese sentences):

(90) ora [**ke**   aSbe   **C**] Suneche **C** (Bangla)
     they who will.come heard
     'Who have they heard will come?' (*matrix*)
     'They have heard who will come' (*embedded*)

If Bangla, like Japanese, places phonological phrase boundaries at Left edges of DPs, then the DP *ke* 'who' will have a phrase boundary to its Left. Consequently, we can use the procedure for creating prosodic *wh*-domains to connect *ke* with either of the two complementizers; the phrase boundary to the left of *ke* will be one edge of the *wh*-domain, and the other boundary will be whatever boundary is present to the right of the relevant complementizer. In other words, the *wh*-phrase in this type of example ought to be able to take scope at either complementizer, as indeed it can.

Next, consider (88), repeated as (91):

(91) ora Suneche **C** [**ke**   aSbe   **C**] (Bangla)
     they heard   who will.come
     'They have heard who will come' (*embedded* only)

If we apply the procedure for creating *wh*-domains to an example like (91), the *wh*-domain will start at the Left edge of *ke* 'who' and extend as far right as necessary to include a complementizer where the *wh*-phrase can take scope. As it happens, there is only one complementizer to the right of *ke* in (91), and hence *ke* can only take scope in this position. If the complementizers in (91) are placed correctly, and if Bangla prosody is like Japanese prosody in this respect, then we have the result we want.

It has in fact been argued that Bangla prosody is not unlike Japanese prosody in the respects that are relevant for this theory. Hayes and Lahiri (1991), Michaels and Nelson (2004), and Selkirk (2006) describe a system for Bangla prosody that typically places prosodic boundaries at Left edges of certain maximal projections; the examples in (92), which would form single phonological phrases, are from Hayes and Lahiri 1991, 87–88:[16]

(92) a. [[ram-er]   Taka] (Bangla)
        Ram-GEN money
        'Ram's money'
     b. [[chobi-r]   jonno]
        pictures-GEN for
        'for pictures'
     c. [[[Tok gur-er]   jonno] durgOndho]
        sour molasses-GEN for   bad.smell
        'the bad smell of sour molasses'

All of these examples have DPs contained in larger structures; in (92a), for example, the DP possessor *ram-er* 'Ram's' is followed by the noun *Taka* 'money'. These DPs are not, however, followed by prosodic breaks, which shows that Right edges of DPs are not typically associated with breaks in prosody.

To see what syntactic boundaries are mapped onto prosodic boundaries, we can consider the phrasing of examples like (93), where the phrasing is indicated by parentheses:

(93) (Omor) (cador)(tara-ke) (dieche) (Bangla)
        Amor   scarf   Tara-DAT gave
        'Amor gave the scarf to Tara'

The phrasing in (93) shows, first, that Bangla is like Basque in that it places a prosodic break just before the verb. Second, we can see in (93) that although Right edges of DPs may not be associated with prosodic breaks (as (92) showed us), Left edges of DPs apparently are.

The second of these is the more important one from our perspective; we are concerned with how edges of maximal projections are typically mapped onto prosodic domains, since it is these maximal projections that may be *wh*-phrases and will therefore potentially determine one edge of a prosodic *wh*-domain. The anomalous prosodic break before the verb makes Bangla resemble Basque (and in fact allows Bangla to join the majority of the languages discussed in this chapter, almost all of which seem to treat the verb as some kind of exception to the general pattern), but this is a red herring, on the theory developed here; the preverbal break will never be one projected by a *wh*-phrase, and hence will never determine the position of such a phrase. This is a good result, since Bangla *wh*-phrases are not constrained, as their Basque counterparts are, to appear in immediately preverbal position[17] (Simpson and Bhattacharya 2003, 137):

(94) jOn **kon boi-Ta**   borders-e   kal   kinlo (Bangla)
     John which book-CLASSIFIER Borders-LOC yesterday bought
     'Which book did John buy yesterday at Borders?'

Thus far, Bangla intonation appears to be relevantly like Japanese intonation, in that Left edges of certain maximal projections (such as DPs) are mapped onto prosodic boundaries. In cases in which a Bangla *wh*-phrase precedes a complementizer, then, Bangla ought to be able, like Japanese, to create a prosodic *wh*-domain extending from the *wh*-phrase to the complementizer, and thus allow the *wh*-phrase to take scope at the complementizer without movement.

In fact, Hayes and Lahiri (1991), Michaels and Nelson (2004), and Selkirk (2006) do discuss data that may support the idea that these Bangla *wh*-questions involve the creation of prosodic *wh*-domains. The main concern of these researchers is actually with focus, rather than with *wh* in situ (though Hayes and Lahiri do suggest that *wh* in situ has the same prosody as focus). What they show is that focus on a particular phrase is characterized by loss of prosodic boundaries after that phrase. Focus has other prosodic consequences, however, which are less congenial to the theory developed here; in particular, a focused phrase is apparently always immediately *followed* by a prosodic break. Given the complicated nature of the data, and the fact that the data are primarily about focus and not *wh*-questions, I will leave this as an issue for future work on Bangla intonation. The point of interest, from our perspective, is simply that Bangla seems to have the right type of prosody to allow *wh*-phrases to remain in situ, just when they precede the complementizer where they take scope. In other words, Bangla has the potential to be just like Japanese, as far as this theory is concerned.

3.5.3 French, Portuguese

French and Portuguese can both form *wh*-questions either via movement or with *wh* in situ:

(95) a. Tu as   vu   qui? (French)
         you have seen who
         'Who did you see?'
     b. Qui tu as vu?

(96) a. O Bill comprou o que? (Portuguese)
         D Bill bought   D what (Pires and Taylor 2007, 2)
         'What did Bill buy?'
     b. O que o Bill comprou?
         D what D Bill bought

Both of these languages have head-initial complementizers:

(97) a. Guillaume croit   [**que** Pierre aime Marie] (French)
         Guillaume believes that Pierre loves Marie
         'Guillaume believes that Pierre loves Marie'
     b. O Pedro disse [**que** ele leu   o quê?] (Portuguese)
         D Pedro said   that he read D what (Pires and Taylor 2007, 9)
         'What did Pedro say that he read?'

Thus, the theory developed here leads us to hope that both languages ought to be metrically like Chicheŵa, as outlined in section 3.3.4—that is, they ought to be languages in which Right edges of maximal projections are typically associated with metrical boundaries. French and Portuguese would then be able to leave *wh* in situ by building a *wh*-domain with one of its boundaries at the right edge of the *wh*-phrase, and its other boundary at the beginning of the clause.

   Both languages have indeed been claimed to impose metrical boundaries at Right edges of maximal projections. Selkirk (1986), for example, makes this claim for French, on the basis of properties of liaison. Liaison causes underlying consonants to appear whenever followed by a vowel-initial word with no intervening prosodic boundary. In (98), for example, the boldfaced consonants undergo liaison and are pronounced:

(98) (Ces très amiable**s** enfants)(e**n**   ont
      these very nice   children of.it have
      avalé) (French)
      swallowed (Selkirk 1986, 395)
      'These very nice children swallowed some of it'

Sandalo and Truckenbrodt (2002) claim that Brazilian Portuguese also
creates prosodic structure by assigning prosodic boundaries at right edges
of maximal projections. The phenomenon they use to study Portuguese
prosody is stress retraction under stress clash: a word with final stress
will retract stress to the penultimate syllable, if the word is followed by a
word with initial stress, and the two words are not separated by prosodic
boundaries. In the following examples, stress is indicated by underlining:

(99) a. (O   café   quente)(queimou a   boca
        the coffee hot   burned   the mouth
        ontem) (Brazilian Portuguese)
        yesterday
        'The hot coffee burned my mouth yesterday'
    b. (O   novo café) (queima a   boca   sempre)
        the new coffee burns   the mouth always
        'The new coffee burns my mouth always'

Thus, stress retraction takes place in (99a), because the head noun and the
following adjective share a prosodic phrase, but not in (99b), in which the
subject and the verb are separated by a phrase boundary.

These two languages are therefore predicted to allow either *wh*-
movement or *wh* in situ, as indeed they do. Particularly for French, there
is clearly much more to say; for instance, French *wh* in situ appears to be
unable to cross tensed clause boundaries (Chang 1997; Bošković 1998;
Reglero 2005):

(100) a. Qu'a   dit Peter que John a   acheté? (French)
          what-has said Peter that John has bought
          'What did Peter say that John bought?'
      b. *Peter a   dit que John a   acheté quoi?
          Peter has said that John has bought what

The same is not true of Portuguese (Pires and Taylor 2007, 10):

(101) O   João e   o   Pedro acham que a   Maria viu
      the João and the Pedro think that the Maria saw
      quem? (Portuguese)
      whom
      'Who do João and Pedro think that Maria saw?'

It is possible that some of the conditions on *wh* in situ in these languages are related to conditions to the prosody of the *wh*-domain (see, in

particular, Cheng and Rooryck 2000 as well as Hamlaoui 2008 for discussion of this possibility for French).

3.5.4 Echo Questions

Many languages with overt *wh*-movement allow *wh* in situ just in the case of "echo questions" like the one in (102b):

(102) a. John bought a motorcycle.
      b. <small>John bought a</small> **WHAT**?

In fact, this exception for echo-questions is extremely widespread, though perhaps not universal. Comorovski (1996) and Bošković (2001) claim that echo questions are impossible in Romanian, for example:

(103) *Ion a   adus   ce? (Romanian)
      Ion has brought what
      'John brought what?'

Still, it is quite common to find *wh* in situ just for echo questions.

In a theory which posits strength and weakness, it is unclear why this should be so. We could, for example, claim that English has an ordinary interrogative complementizer with a strong feature, and an "echo question" complementizer with a weak feature. While this would get the English fact, it seems to miss the point, which is that (a) the English pattern is extremely common, and (b) the reverse is (as far as I know) unattested; there are no *wh* in situ languages that require movement just for echo questions. A more interesting tack might be to claim that echo questions lack complementizers at all, and hence lack a Probe for the *wh*-phrase; this approach would be left with the burden of explaining how such questions can be interpreted as questions.

The theory developed here allows us to make another kind of move. Echo questions are typically questions in which all the material, apart from the *wh*-phrase itself, is old information; in fact, this material is the limiting case of old information, in the sense that it is typically a repetition, sometimes with slight rephrasing, of something previously said. This results in the destressing of all the non-*wh* material, indicated in (103b) by small type.[18] This chapter has proposed that *wh*-phrases must be separated from the complementizer at which they take scope by as few prosodic boundaries (of a certain type) as possible. The bulk of the chapter has concentrated on two main strategies for achieving this: movement of the *wh*-phrase closer to the complementizer, and creation of a new Minor Phrase, making use of existing boundaries.

In an example like (103b), we might be seeing a third kind of strategy. It is possible, at least, that the destressing of all the non-*wh*-material in the sentence reflects a lack of prosodic structure; this old information, on this type of account, is not assigned the type of prosodic structure that it would be if it were new information. Consequently, (103b) might be a prosodically well-formed *wh*-question to begin with; there might be no offending prosodic boundaries between the *wh*-phrase and the complementizer.

At the moment, this is speculation; I have no facts to offer about the prosody of echo questions that would support this. The account does have the virtue, however, of explaining why echo questions are associated with *wh* in situ, and why there are no languages in which echo questions are associated with movement not found in non–echo questions. The properties of echo questions, on this type of account, are linked to the fact that old information tends (universally?) to be prosodically bleached, and hence to make the creation of a prosodically well-formed *wh*-question easier, even in languages that must normally resort to *wh*-movement.

3.6 Conclusion

I began this chapter by noting that languages seem to differ in how they form their *wh*-questions. The chapter has been an exploration of the idea that this is in fact false. Languages do not vary in how they form their *wh*-questions; in every language, *wh*-questions are formed by arranging for the *wh*-phrase and the complementizer associated with it to be separated by as few (Minor) phrase boundaries as possible, for some level of Minor phrasing.

How this universal goal is achieved, of course, is in fact a matter of crosslinguistic variation. But the crosslinguistic variation appears to follow, once the universal goal is stated in this way; languages treat their *wh*-phrases differently because their complementizers are in different places, and because the basic rules for how prosodic structures are formed can differ from language to language. Languages (like Japanese, Chichewa, and possibly Spanish, French, Portuguese, and Bangla) that place complementizers on one side of *wh*-phrases and habitually map the other side onto prosodic boundaries are able to satisfy the prosodic conditions on *wh*-questions without movement, and hence can leave *wh*-phrases in situ. Languages that place the complementizer and the prosodic boundaries on the same side of maximal projections (such as Basque, Tagalog, and hopefully English) cannot directly manipulate the prosody in this way, and must resort to movement, doing everything possible to bring the *wh*-phrase and the complementizer closer together.

Much work remains to be done, of course. The theory will need to be tested on many more languages than the handful that I have managed to apply it to here; for many languages, this will require study of the basic mapping of syntax onto prosody. I have not attempted to apply this theory to multiple-*wh* questions, another domain of crosslinguistic variation, which needs further study.

Ultimately, the hope is to apply this way of thinking to other types of movement, as well. We have grown accustomed to being able to stipulate that this or that type of movement (not only *wh*-movement, but also scrambling, head movement of the verb to T, and so on) is present or absent in a given language. The idea here has been to derive this apparent parameter from other parameters, just in the case of *wh*-movement. If this attempt proves successful, then we need to undertake the same project for other types of movement. In general, the goal should simply be the general one of linguistics: to describe languages in such a way that, properly understood, they do not differ, at least not in as many ways as one might at first think.

At the beginning of this chapter I raised a "look-ahead" problem associated with this theory: what is the right way of understanding the interactions between the syntax and the phonology, such that conditions on the prosody dictate how the syntax is to treat *wh*-phrases? In my opinion, it is probably too early to try to answer this question, though it is sure to be a pressing one if the research program outlined here continues. One important part of seeking a solution will be developing a better understanding of what types of phonological information the syntax is allowed to "respond" to. I have been claiming in this chapter that syntactic operations can in part be triggered by considerations of prosody, but I suspect that not all properties of phonology can influence the syntax in this way; we will probably not find movements triggered by the need to put onsetless syllables in positions where they can acquire onsets from preceding consonant-final words, for example. If this turns out to be correct, then we probably do not want to solve the "look-ahead" problem by brute force, allowing the syntax full access to the information in the phonological representation.

In section 3.3.3.2, I briefly addressed the question of how we ought to choose the particular maximal projections that are mapped onto prosodic boundaries. I suggested there that the maximal projections that are relevant for prosody might be specifically the phases. The example under consideration there was Tagalog; we saw evidence that KPs are phases in Tagalog, and I showed that an algorithm mapping left edges of KPs onto Minor Phrase boundaries, together with deletion of the boundary imme- diately following the verb, could account for the Tagalog data.

In chapter 2, it was important that Spell-Out take place along the lines developed by Nissenbaum (2000); when a phase is completed, all the ma- terial in the phase except for the edge is sent to PF, while the material at the edge remains part of the syntactic computation. Suppose this process of Spell-Out is also responsible for inserting prosodic boundaries (or ob- jects which will be interpreted as prosodic boundaries by PF, at any rate) on the right or left edge of the phase. Consider the derivation of a Taga- log sentence like the one in (104):

(104) Nakita nila   ang lola   ni Maria (Tagalog)
      ACC-saw NG.they ANG grandmother NG Maria
      'They saw Maria's grandmother'

The derivation of (104) might begin by constructing the KP *ni Maria* 'NG Maria':

(105)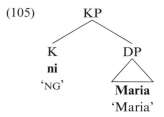

Once this KP is fully constructed, it would undergo Spell-Out, which would have two consequences. Spell-Out would send the DP *Maria* to PF, where it would ultimately be converted into a string of phonemes, and it would also assign a prosodic boundary to the left edge of the KP phase:

(106)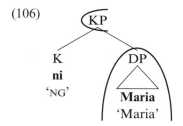

The syntactic derivation would continue, Merging the KP in (99) as the possessor of a larger DP:

(107)

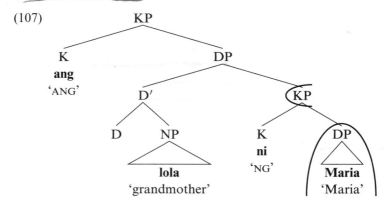

The previous Spell-Out operation makes the structure linearizable; the DP *Maria* is not linearized in the same phase as the DP *lola* 'grandmother', so the Distinctness principle of the previous chapter is respected. Now that the larger KP has been constructed, Spell-Out can apply again:

(108)

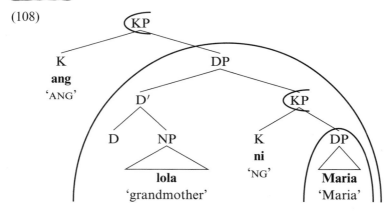

In the approach to Spell-Out developed here, the edge of the phase is not completely unaffected by Spell-Out; rather, the PF component returns to the syntax an object that has been partly annotated for prosody. The proposal resembles Fox and Pesetsky's (2004) idea of Cyclic Linearization, in which the syntax manipulates objects which are occasionally annotated for word order.

In the derivation sketched above, the syntax is not simply granted full understanding of the PF representation; rather, certain aspects of the PF representation are determined via Spell-Out before the syntactic derivation has ended. This is probably a desirable result; as I remarked earlier, there are aspects of phonology to which syntax is probably never sensitive, such as syllable structure. As this research program progresses, we may hope to shed further light on the nature and extent of syntax's understanding of phonology.

4   Conclusion

Existing syntactic theories offer explanations of many of the phenomena treated in the preceding chapters. The contrast between overt and covert *wh*-movement has been a locus of very fruitful work, but this work has mainly focused on the syntactic and semantic properties of these dependencies, with relatively little attention devoted to why languages vary in this way. Existing Case theory successfully constrains the distribution of DPs, but does so in a fairly arbitrary way; DPs (but not other kinds of phrases) need Case, and certain heads (but not others) are capable of supplying it, for reasons that are not well understood.

The purpose of the work presented here has been to deepen these explanations, as well as to broaden their empirical coverage. Case, on the theory developed in chapter 2, is a particular instance of a more general condition on syntactic structures, resulting from a ban on linearization statements that appear to linearize a given node with itself. The ban on self-linearization is not restricted to DPs; they simply offer one very salient example of it. Moreover, the ban on self-linearization, unlike a requirement that DPs receive Case, seems susceptible to explanation in more general terms; no object can enter into an asymmetric relation (such as linear precedence) with itself, and representations that seem to require this are uninterpretable. Chapter 3 attempted to derive the behavior of *wh*-phrases without using diacritics such as strong and weak features; the relevant parameters distinguishing between languages, on that theory, are previously established ones having to do with the position of complementizers and the organization of prosody more generally.

If the accounts developed here are correct, we are forced to certain conclusions about the architecture of the grammar more generally. In particular, I have claimed in several places that the construction of phonological representations begins earlier in the syntactic derivation than was previously thought. In sections 2.4.3 and 2.4.4, I claimed that the

grammar apparently seeks to avoid Distinctness violations throughout
the derivation; the conditions that Distinctness imposes on word order
are apparently relevant even before the syntax hands the structure to phonology. In section 3.6, I suggested that multiple Spell-Out might solve the
problems of look-ahead posed by the proposal of that chapter; Spell-Out
of a phase might return to the syntax an object annotated for prosodic
structure (and perhaps word order, following the suggestion of Fox and
Pesetsky 2004).

The theories presented here raise many questions for possible further
study. Like any explanations, my explanations have to stop somewhere,
and we should ask whether still deeper levels of explanation are possible.
Why should a *wh*-phrase and its scope position have to share a prosodic
domain? Can any of the parameters assumed here (including the distinction between initial and final complementizers, and the different levels of
Distinctness among types of DPs) be made to follow from more general
principles? The theory of Distinctness allows for a number of ways of
fixing potentially offending structures; what determines which method is
used in a given language? The theory developed in chapter 3 attempts to
predict whether a language will have overt *wh*-movement or not; can we
replace all arbitrary statements about which movements are possible in a
given language with explanations of this kind? Having done so, can we
develop a theory from which such explanations can be derived?

The theme of this book has been the conversion of syntactic trees into
phonological representations. Our knowledge of the conditions on that
conversion is still underdeveloped, but seems to me to hold great promise
for our future understanding of the nature of syntax.

Notes

Chapter 2

1. For other work on a "syntactic OCP," see Aissen 1979, Hoekstra 1984, Mohanan 1994b, Yip 1998, Anttila and Fong 2000, and the references cited there. Ultimately, we may want a unified OCP for both phonology and syntax; assuming that phonological representations also involve linearization statements (Raimy 2000), then a version of the account developed here might be generalized to phonology.

2. It will be crucial in what follows that only *transitive* vP is associated with a phase boundary—that is, I will have to side with Chomsky (2000, 2001) and den Dikken (2006a) and against Legate (2003) in making intransitive vP a nonphase.

3. Rehearsing the arguments for this conclusion would take us too far afield; see Halle 1990, Embick 2000, and Embick and Noyer 2006 for some discussion.

4. This raises the natural question of what happens when multiple identical lexical heads appear in a Spell-Out domain. I will defer discussion of this question until note 34, when we will see a relevant case. What we will see there, distressingly for the theory offered here, is that such examples are well formed.

5. A reviewer suggests that the difficulty with the sentences in (7) might have to do with stress clash; the idea is that in (7a), for example, *Mary* receives nuclear stress, and is therefore unacceptably close to *John*, which is also stressed. The reviewer suggests testing examples like (i) below, which should avoid this problem:

(i) *Every man admired every woman, except [John] [the woman in the kitchen]

Since nuclear stress is on *kitchen* in this example, the reviewer suggests that it should avoid the stress clash problem. In fact, the example is just as bad as (7a), so it appears that we cannot appeal to stress clash to explain the facts in (7).

6. For a similar argument regarding restructuring, see Wurmbrand 1998, 2003.

7. A reviewer wonders whether a version of the EPP could be made to follow from this reasoning; the subject must move to the specifier of TP in order to avoid being linearized with other DPs. While this approach could certainly work in sentences with transitive verbs, the existence of EPP effects in intransitive clauses could not be made to follow in this way, as far as I can see; some version of the EPP is apparently still required.

8. Such bans have indeed been proposed, and it seems reasonable to try to relate them to the facts I am trying to capture with Distinctness; since they often do seem to be crucially concerned with the phonological shape of the adjacent items, perhaps they are the result of Distinctness continuing to apply after lexical insertion has taken place. See Perlmutter 1971, George 1980, Menn and MacWhinney 1984, Kornfilt 1986, Grimshaw 1997, Bošković 2001, Ackema 2001, Neeleman and van de Koot 2005, and Biberauer 2008 for relevant discussion.

9. The same reasoning yields a new account of *wanna*-contraction; *wanna*-contraction is possible just in case *want* and *to* are both linearized in the same phase. See Ausín 2000.

10. Examples like (i) raise further complexities:

(i) Into the room are walking singing men.

(p. 13)

We already know, from section 2.1.5, that the inverted subject in such an example must be spelled out with the lower phase. As long as v_C is present and acts as a phase head in this example, the account in the text can deal with the well-formedness of (i); *walking* can move to the edge of the v_CP phase and be spelled out with the higher phase, while *singing men* is spelled out with the lower phase. In the current section, it is important that v_C is a phase head only when the verb is transitive, but I have had nothing to say about why that is the case. Perhaps v_C must also be present in order to move the locative expression past the subject.

11. If Italian raising infinitives are not phases, another problem arises for well-formed examples like (i):

(i) Paolo sembra dormire (Italian)
Paolo seems   sleep-INF
'Paolo seems to be sleeping'

Why do the *v* associated with *sembra* and the *v* associated with *dormire* not trigger a Distinctness violation? A possible answer is that Italian (unlike English) has head movement of V to T, so that *v* is a subpart of a larger, complex head. The effects of head movement will have to be calculated carefully, however, since head movement of *v* to v_C apparently fails to protect it in example (35).

12. Many thanks to Degif Petros, the source of all the Chaha data.

13. See sections 2.4.1.1 and 2.4.3 for discussion of why animacy plays a role in these examples.   *(p. 55)* *(p. 26)*

14. We will see a similar need for head-complement ordering in the next chapter.

15. If DP is the complement of K, we must assume that linearization is able to distinguish between a spelled-out DP and an un-spelled-out DP.

16. In Hindi, the marker can apparently also appear in inanimate objects, as long as they are definite (while the crucial property for inanimates is specificity; Mohanan 1994a, 79–80).

17. Thanks to an anonymous reviewer for pointing this out to me.

18. In both Hindi and Spanish, we are led to wonder why it must be the direct object and not the indirect object that shifts into the higher phase. One possibility is that the K heads *-ko* and *a* have another life as adpositions. The adpositional versions of these markers will have to be indistinguishable from the K versions for purposes of linearization, but would render the indirect object a PP, perhaps disqualifying it for object shift.

19. Given the claim in the introduction, supported in section 2.2.1.1, that even phonologically null elements must be linearized, it may be unclear why making the relative operator null obviates the Distinctness violation. As I also remarked in the introduction, however, if the syntactic derivation is itself responsible for making some syntactic object phonologically null, that object need not be linearized (the example I offered was copies left behind by movement, which cannot participate in linearization if the account developed here is correct). The null relative operator, then, will have to be the consequence of an actual deletion operation in the syntax, rather than simply being a phonologically null lexical item.

20. Pesetsky and Torrego (2006) offer a different account; they argue that *whom* in examples like (72a) is in fact an agreeing complementizer, rather than an operator. As they note, the challenge for them is then to account for the well-formedness of examples like (72b). They suggest that the problem is more general, with examples like (69a) also being less than completely ill-formed for many speakers (their example parallel to (69a) is given in (i)):

(i) % a person [whose virtues] to admire

21. For this account to succeed, the CPs cannot be separated by a Spell-Out boundary.

22. I have represented topicalization here as movement to the specifier of CP; as a reviewer points out, there are indeed approaches to topicalization (such as that of Rizzi 1997) that introduce a new head (Top) hosting movement to its specifier. As long as TopP is not a phase, the conclusions of this section are unharmed (and see section 2.2.2.2 for an argument that requires us to deny phasehood to TopP). The TopP projection is standardly claimed to be a subpart of the "CP Layer"; to say that "CP is a phase," in this type of approach, is presumably to say that some privileged head of the CP layer is a phase (and if TopP is not always present, that privileged head should not be Top).

23. As Kazuko Yatsushiro (personal communication) points out, the contrast in (89) is not simply a confusion about which of the two *ga*-marked DPs is the experiencer and which is the target of the emotion. Adding *no koto* to one of the DPs marks it unambiguously as the target of emotion, but there is still a contrast between (i) and (ii):

(i) *[Sensei-no   koto]-ga   suki na gakusei-ga   koko-ni oozei iru kedo, **dono**
     teacher-GEN matter-NOM like   student-NOM here-DAT many be but   which
     gakusei-ga dono sensei-no   koto-ga   ka oboeteinai. (Japanese)
     student-NOM which teacher-GEN matter-NOM Q remember-NEG
     'There are lots of students here who like teachers, but I don't remember which student
     which teacher'

(ii) [Sensei-no   koto]-o   suki na gakusei-ga   koko-ni oozei iru kedo, **dono**
     teacher-GEN matter-ACC like   student-NOM here-DAT many be but   which
     gakusei-ga dono sensei-no   koto-o   ka oboeteinai.
     student-NOM which teacher-GEN matter-ACC Q remember-NEG
     'There are lots of students here who like teachers, but I don't remember which student
     which teacher'

24. The reviewer wonders whether the lack of multiple clefting in English could be handled by Distinctness:

(i) *It's [John Mary] that likes

Since we have already seen in this section that English differs from Japanese in always treating DPs alike for Distinctness, the reviewer suggests, perhaps we could rule out (i) as a Distinctness violation. Unfortunately, as the reviewer notes, this idea predicts that English multiple clefting will become possible if the clefted elements are not both DPs. This appears not to be true:

(ii) *It's [John with Mary] that danced

Consequently, I think we must look elsewhere for an explanation of the ban on multiple clefting in English; multiple clefting and multiple sluicing, for example, do not exhibit the same behavior.

25. Given this conclusion, we might wonder how Japanese allows multiple-nominative examples like (i):

(i) John-ga   Mary-ga   suki da (Japanese)
    John-NOM Mary-NOM like COP
    'John likes Mary'

See Yang 2005 for arguments that the Korean version of this construction always involves A-bar movement of the higher nominative DP to a higher position, perhaps introducing a phase boundary between the two DPs.

26. Many thanks to Kleanthes Grohmann, Winnie Lechner, Joachim Sabel, Uli Sauerland, Michael Wagner, and Susi Wurmbrand for their judgments.

27. Many thanks to Sabine Iatridou for pointing this out to me.

28. Many thanks to Martina Gračanin-Yüksek for pointing this out to me, and for much subsequent discussion.

29. In (106a), *kojeg dječaka* 'which boy' exhibits syncretism for case, and hence could be labeled either genitive or accusative; I have used GEN here and in what follows.

30. Many thanks to Alya Asarina, Lydia Grebenyova, and Arthur Stepanov for their judgments.

31. We might also expect to be able to save the structure by moving the embedded DP to a high position within the embedding DP, raising it out of the c-command domain of heads that would cause linearization problems. The general prediction is that movement of this type will be in complementary distribution with the kind of generalized differential case marking under discussion here.

32. We also predict that this kind of gerund must lack a PRO subject, since the DP PRO and the D of the gerund would be unlinearizable. Some evidence for this conclusion comes from the apparent lack of a binder for anaphors in these gerunds (i–ii):

(i)   [PRO$_i$ killing yourself$_i$/each other$_i$] isn't the answer

(ii) *[the killing of yourself$_i$/each other$_i$] isn't the answer

If the gerund in (ii) is to contain a PRO$_{arb}$ subject, then PRO$_{arb}$ will have to be something less than a full DP, on this account.

33. Given that these examples involve passives, we must apparently conclude that it is possible to passivize across a phase boundary, if we want to maintain the assumption that Distinctness is crucially sensitive to phase boundaries. The fact in (135) might be related to a puzzle that came up in section 2.2.1.1; the double-infinitive filter appears in some Romance languages, like Italian, but not in others, like Spanish and French. Perhaps the relevant difference between Italian and the other Romance languages has to do with whether the infinitival verbs in question are relevantly like English bare infinitives or like English infinitives with *to*.

34. In (145b), as I mentioned earlier, there are two instances of the lexical head N (*beyt* 'house' and *mora* 'teacher'). It is comparatively easy to construct well-formed examples in which these two instances of N are identical (Iatridou and Zeijlstra 2009):

(i)   morat morat ha-mora (Hebrew)
      teacher teacher the-teacher
      'the teacher's teacher's teacher'

In the introduction, I suggested that lexical heads might be immune to Distinctness effects because they undergo early lexical insertion, which makes it easier for the grammar to distinguish between instances of (for example) N. Examples like (i) are certainly problematic for such an explanation. One possibility is that early insertion makes it possible for the grammar to see that the multiple instances of N are in fact the same, so that accurate linearization of them with respect to each other is unnecessary; with functional heads, by contrast, because the heads undergo late insertion, the grammar cannot know whether they are identical or not, and must attempt to linearize them.

35. Thanks to Ken Hale and Andrew Carnie for help with Irish facts.

36. As a reviewer notes, this would predict that Irish ought to be like English, and not like German, with respect to the phenomena discussed in section 2.3.2. That is, Irish ought to ban multiple sluicing with DP remnants, and also ought to ban multiple DPs within the *v*P, since all Irish DPs would be alike for purposes of linearization. As the reviewer also notes, this second prediction has been argued to hold; Bobaljik and Carnie (1996) and McCloskey (1996) argue that the subject has raised out of the *v*P in Irish, despite the VSO word order. For instance, McCloskey (1996, 269) notes that the subject must precede adverbs like *go minic* 'often':

(i) Chuala Róise go minic roimhe   an t-amrán sin
    heard   Róise often   before-it that-song
    'Róise had often heard that song before'

37. Surprisingly, Classical Arabic has a form of construct state in which determiners are removed but case remains (Walter 2005, 12):

(i) kitaab-**u**   l-walad-i (Classical Arabic)
    book-**NOM** the-boy-GEN
    'the boy's book'

Classical Arabic differs from Irish in that it does distinguish nominative and accusative case; perhaps we can take this as evidence that Classical Arabic K is a phase head.

38. Salanova entertains the possibility that languages might vary in how they treat the edge of the highest phase. Another possibility, as he notes, is that the treatment of phases is uniform but that the landing site of *wh*-movement in RP Spanish is not actually the highest specifier of the CP phase. The issue is of considerable interest, since we need some way of guaranteeing that not all languages encounter the difficulties that RP Spanish does in linearizing its *wh*-questions—for example, English questions will have to be different, perhaps in one of the ways outlined here.

39. I have changed Harbour's presentation mainly in order to give the feature [animate] more prominence in the discussion, and also to make the complement of the ϕ-head a functional projection rather than a lexical one.

40. Roughly, Harbour's idea is that in languages exhibiting Person-based splits for ergativity, ergative case is associated specifically with the "extra" features that are present simply to satisfy the selectional requirement imposed by the *v* head. Because first- and second-person arguments are universally required to bear [±author, ±participant] features, these features are never just inserted to satisfy the requirement imposed by *v*; as a result, if any type of nominal avoids ergative marking, it is the first- and second-person pronouns. Third-person arguments, by contrast, might or might not be inherently specified for [±author, ±participant], depending on the particular language in question. If these features are not inherently present, they must be added for third-person arguments to appear in the specifier of *v*P, and these added features trigger the emergence of ergative case in some languages. The observation that languages with Person-based splits for ergativity may mark third-person subjects ergative and first- and second-person subjects nominative, but never the other way around, is part of the Silverstein hierarchy.

41. See also Skopeteas and Verhoeven (2009), who develop an account of word-order tendencies in Yucatec Maya in terms of Distinctness.

42. Coon has in fact recorded spontaneous instances of sentences with multiple postverbal DPs, but in elicitation contexts they are apparently typically rejected. She notes (personal communication) that she would be surprised to hear examples like (184), in which both DPs have the D *jiñi*; her examples of multiple postverbal DPs have involved one instance of *jiñi*, and another of the determiner for proper names, *aj*:

(i) Mi i-   jats' **jiñi** gringo **aj**- Pedro (Chol)
    PERF 3ERG hit   D   gringo D   Pedro
    'The gringos will hit Pedro'

43. Nothing about the account developed here would explain why NP complements in some language (e.g., Mohawk) must incorporate. Massam's (2001) and Coon's (forthcoming) work suggests that this is a language-specific choice, because languages can in fact have NP complements that do not incorporate (so-called pseudoincorporation).

44. *liang ci* may also appear at the end of the sentence; this option is apparently somewhat degraded in the applicative, for reasons I do not understand.

45. A reviewer notes another condition on movement past the particle, which is that it can apparently only apply to one phrase; both of the examples in (i) are ill-formed:

(i) a. *The secretary sent [the stockholders] [a schedule] **out**
    b. *I sent [a schedule] [to the stockholders] **out**

The facts in (i) must not be Distinctness effects, since the labels of the moved items are clearly irrelevant here.

46. I am very grateful to Pierre Mujomba for his patient work with me on his language.

47. In fact, some of the work on Kinande argues that the preverbal subject may occupy any of several distinct positions (for arguments to this effect, see Schneider-Zioga 2007 and Miyagawa, forthcoming). I will abstract away from such variation in what follows; for the accounts developed below to succeed, all the preverbal subject positions in Kinande will have to be separated from postverbal material by a Spell-Out boundary.

48. Baker and Collins (2006) have a pair of examples (their example (15)) in which the subject cannot follow the linker; in their examples, the other VP-internal phrase is an instrumental, and the linker cannot appear at all:

(i) a. ?Olukwi si-lu-li-seny-a   bakali (*b') omo   mbasa
       11-wood NEG-11s-PRES-chop-FV 2-women 2.LI 18-with 9-axe
       'WOMEN do not chop wood with an axe' (Kinande)
    b. *Olukwi si-lu-li-seny-a   omo   mbasa (mo) bakali
       11-wood NEG-11s-PRES-chop-FV 18-with 9-axe   18.LI 2-women

49. It may be relevant, for example, that the examples in (203) and (204) involve instrumental expressions, while the one in (205) contains a locative. It will be crucial in what follows, however, that even locative expressions can be represented as either PP or DP.

50. Baker and Collins (2006) offer a structurally analogous set of sentences (their example (45), repeated below as (i)), and annotate the third with (?). I do not know whether this represents a real difference between speakers; if it does, I will have to leave the judgments in (i) for future research:

(i) a. N-a-seny-er-a   omwami **y'** olukwi   omombasa (Kinande)
       1ss-T-chop-APPL-FV 1-chief   1.LI 11-wood 18-with-axe
       'I chopped the wood for the chief with an axe'
    b. N-a-seny-er-a   olukwi   **l'** omwami omombasa
       1ss-T-chop-APPL-FV 11-wood 11.LI 1-chief   18-with-axe
    c. (?)N-a-seny-er-a   omombasa **m'** omwami olukwi
       1ss-T-chop-APPL-FV 18-with-axe 18.LI 1-chief   11-wood

51. None of the other logically possible orders of postverbal phrases are grammatical either. As Baker and Collins (2006) argue, Kinande postverbal word order is fixed, apart from the freedom created by the ability to move one DP to the prelinker position.

52. Later Spell-Out of the v_CP phase will separate the subject (the null pro, referring to 'you'), which will be at the edge of this phase, and the benefactive object (*ómwami* '1-chief').

53. These *wh*-questions involve clefts, and I assume that the *wh*-phrase and the subject are linearizable by virtue of being in separate clauses. Kinande does also have a noncleft strategy for *wh*-movement (see Schneider-Zioga 2007 for much discussion); even in this strategy, the fronted *wh*-phrase is followed by a complementizer, which I will take to be a phase head.

54. For example, an utterance-final verb without lexical tone typically bears an utterance-final low tone on its last mora, and a phrase-final high tone on its penultimate mora (Jones and Coon 2008):

(i) eri-húm-à (Kinande)
    INF-hit-FV
    'to hit'

Jones and Coon (2008) argue that the phrase-final high tone must be anchored to the first consonant in the macrostem. For monosyllabic verb stems that begin with vowels, this means that the phrasal high tone is realized after the stem, combining with the utterance-final low tone to yield a final contour tone:

(ii) ery-es-â (Kinande)
 INF-play-FV
 'to play'

In (ii), the phrasal high tone is realized after the /s/ in the verb stem -es- 'play', since this is the first consonant of the macrostem. If we add an object agreement marker, the high tone will have an earlier consonant to anchor to, and hence will move to an earlier position in the verb:

(iii) eri-mw-és-à (Kinande)
 INF-**1**.O-play-FV
 'to play him'

55. Many thanks to Victor Manfredi and Patrick Jones for helpful discussion of Kinande morphology; neither should be held responsible for the conclusions I have drawn here, of course.

56. Baker and Collins (2006) have a pair of examples (their example (15)) in which the subject cannot follow the linker; in their examples, the other VP-internal phrase is an instrumental:

(i) a. ?Olukwi si-lu-li-seny-a   bakali (*b') omo   mbasa
  11-wood NEG-11S-PRES-chop-FV 2-women 2.LI 18-with 9-axe
  'WOMEN do not chop wood with an axe' (Kinande)
 b. *Olukwi si-lu-li-seny-a   omo   mbasa (mo) bakali
  11-wood NEG-11S-PRES-chop-FV 18-with 9-axe   18.LI 2-women

This may have a bearing on the question (to be discussed shortly) of why the linker cannot rescue Distinctness violations involving postverbal subjects; perhaps the linker is simply prevented from appearing in certain instances of subject-object reversal, for reasons still to be explained.

57. The creation of multiple specifiers via "tucking in" (Richards 1999, 2001) will be problematic for this approach; tucking in creates the highest specifier first, but it is the highest specifier that is treated as highest for purposes of later calculations of locality (see Richards 2001 for arguments).

58. Though the tree in (258) depicts movement of the applied object to the specifier of v_C, it is in fact possible for either internal argument to move to the prelinker position, as we have already seen:

(i) a. Omúlumy' a-ámá-hek-er'   omúkalí y' akatébé (Kinande)
  1-man   1S-PRES-carry-APPL 1-woman 1.LI 12-bucket
  'The man carries the bucket for the woman'
 b. Omúlumy' a-ámá-hek-er'   akatébé k' omúkali
  1-man   1S-PRES-carry-APPL 12-bucket 12.LI 1-woman
  'The man carries the bucket for the woman'

Thus, v_C must first Agree with the subject, but can then freely Agree with either of the two remaining DPs. The behavior of v_C in this regard is reminiscent of the behavior of Bulgarian interrogative C, which must Agree first with the highest wh-phrase but can then Agree with any wh-phrase below it (a fact noted by Rudin 1985, and subsequently discussed by Bošković 1997 and Richards 2001):

(ii) a. Koj kakvo na kogo e   dal?
  who what to whom AUX given
  'Who gave what to whom?'
 b. Koj na kogo kakvo e dal?

We might hope to derive the locality conditions on movement of DPs in Kinande, then, from general conditions on locality. Subject-Object Reversal is possible because removal of the subject's augment disqualifies it as a Goal (see the discussion of (251) in the main text), and either of the two internal arguments may move to the specifier of v_C, even if both internal arguments retain their augments, because v_C has already obeyed general conditions on locality by Agreeing first with the subject. In terms of the theory developed in the text, we might take these facts as evidence that the grammar does not keep a perfect record of the order in which problematic elements were Merged; it is aware of which uninteptretable feature was Merged most recently, but its grasp of derivational history is no more detailed than that.

59. I should also admit that I have no idea how to capture the special status of nonargument accusatives in causative constructions, which yield only the "weak" kind of Distinctness violation that is subject to repair (examples (246a,b), repeated here as (i) and (ii)):

(i) Hanako-ga   Taroo-**ni**/*-**o**   hamabe-o hasiraseru (Japanese)
     Hanako-NOM Taroo-DAT/ACC beach-ACC run-CAUS
     'Hanako makes Taroo run on the beach'

(ii) [Hanako-ga   Taroo-**o**   hasiraseta   no wa] hamabe-**o** da
     Hanako-NOM Taroo-ACC run-CAUS-PAST C TOP beach-ACC is
     'What Hanako made Taroo run on is a beach'

Perhaps expressions like *hamabe-o* 'beach-ACC' are marginally reinterpretable as PPs, rendering them linearizable (though for numerous arguments against such a proposal, see Poser 2002 and the references cited there).

60. On the other hand, I do not know of any languages in which relative-clause operators are rescued by giving them genitive case. If this is a real gap, it needs an explanation. One explanation would keep at least some aspects of the mechanics of Case checking as developed in Minimalism, ruling out genitive Case assignment to an A-bar moved phrase, via whatever mechanism rules out improper movement.

61. For languages that only make one argument into a KP, the choice of which argument to make into a KP could be forced in our theory by allowing languages to choose whether K is a phase head or not. The K in (277a) must be a phase head, spelling out its complement DP in a separate Spell-Out domain. If this instance of K were not a phase head, linearization would still fail, since there would still be an instance of DP c-commanding DP within the XP phase. Languages that mark the lower of the two DPs (that is, nominative-accusative languages), then, must treat K as a phase head if it is to help with linearization. A K that was not a phase head could still save a structure in which one DP c-commands another, however, as long as it was the higher DP that was made into a KP. The resulting structure would have a KP c-commanding a lower DP, which would be a linearizable structure; here KP need not be a phase head, since the DP dominated by KP is safe from being c-commanded by the other DP. Languages in which K was not a phase head, then, would be forced to be ergative, making the subject rather than the object into a KP, if linearization is to be achieved via insertion of K. On the other hand, I will shortly discuss some Dutch facts which can be characterized using a K which is not a phase head.

62. This raises the question, in this framework, of why English has case morphology at all. Perhaps the answer is a diachronic one; the case morphology that we see in English now is useless for linearization, but is descended from case morphology in older stages of English, which was capable of alleviating Distinctness violations.

63. The Dutch facts are structurally parallel to the Spanish facts discussed in section 2.4.1.1:

(i) a. Aman   (* a) el dinero (Spanish)
         they.love   *a* the money
         'They love money'
     b. Su amor al   dinero
         his love *a*-the money
         'his love of money'

Spanish inanimate nominals can be distinguished from animate ones that c-command them (as in (ia)), but a nominal complement of a nominal must always be protected by insertion of the phase head *a*. Similarly, Dutch nominals can be linearized together with c-commanding nominals (as in the sluicing example in (281b)), but not with dominating nominals (as in the nominal complement example in (282b)).

64. Proper names of people and animals have a different set of markers: the proper-name versions of *ang*, *sa*, and *ng* are *si*, *kay*, and *ni*, respectively.

65. In some Algonquian languages (such as Wampanoag) obviation is limited to animate arguments.

66. Another way of achieving this distinction would follow from the work of M. Richards (2004, 2007). Taking as his starting point Chomsky's (2005) idea that non–phase heads like T inherit their features from phase heads like C, M. Richards (2007) derives the result that phase heads must have non–phase heads as their complements. This would correctly rule out a tree like (292). On the other hand, it is not clear how the account could be generalized to the other facts in this chapter.

67. It is tempting to try to account for the failure of Case Resistance with interrogative clauses in terms of Distinctness; we might say, for example, that the *wh*-phrase intervenes between the higher P and the lower C. Such an account would be difficult to reconcile with the theory as it has been developed so far, however; as Distinctness is currently formalized, a specifier between two heads does not relevantly intervene between them (see section 2.2.2 for evidence for this conclusion). The requirement of *of*-insertion in (296) would also be mysterious for such an account (it does not seem to be related to an interaction between the D-feature of the *wh*-phrase and the D of the higher NP, since even nonnominal *wh*-phrases, which presumably lack a D, require *of* here).

Chapter 3

1. We encounter a similar type of look-ahead problem in Fox's (2000) account of QR, which establishes (quite convincingly) that the syntactic operation of QR is sensitive to the possible semantic consequences, occurring only when it would meaningfully alter the LF of the sentence. Taken together, the look-ahead problems suggest that our understanding of the interfaces is flawed in some way.

2. The dip in the middle of the pitch track for *Kyooto* in (6b) is associated with the voiceless stop in the middle of the word.

3. For discussion of tonal patterns in *wh*-domains in other dialects of Japanese, see Igarashi and Kitagawa 2007 and the references cited there.

4. For other approaches to the Japanese *wh*-intonation facts, see Pierrehumbert and Beckman 1988, Truckenbrodt 1995, Ishihara 2007, and the references cited there.

5. As a reviewer notes, this algorithm will have to be constrained by a general principle to the effect that *wh*-domains should be no larger than they must be in order to satisfy the prosodic conditions on *wh*-questions.

6. This is hopefully a special instance of a more wide-ranging constraint. For similar proposals, see Blaszczak and Gärtner 2005 as well as Hirotani 2005.

7. As a reviewer notes, this prediction requires me to disagree with Takahashi's (1993) proposal that long-distance scrambling of *wh*-phrases in Japanese can be equated with overt *wh*-movement.

8. For some discussion of related phenomena, see Seidl 2001, McGinnis 2001, and Ishihara 2001.

9. In fact, what Selkirk and Tateishi say is that Left edges of maximal projections are mapped onto *Major* Phrase boundaries. Since Major Phrases are always completely decomposed into Minor Phrases, this has the effect that every edge of a Major Phrase boundary is also the edge of a Minor Phrase boundary, which makes the statement in the text an

accurate one. I am concentrating on the distribution of Minor Phrases because these are the phrases whose construction differs in theoretically useful ways between Japanese and Basque.

10. It would be comforting to have a less Basque-specific answer to this question, however, since rightward *wh*-movement seems to be crosslinguistically very uncommon, at least in spoken languages; see Neidle et al. 2000 as well as Cecchetto, Geraci, and Zucchi 2007 for arguments that rightward *wh*-movement is indeed found in American Sign Language and Italian Sign Language, and perhaps in signed languages more generally. More research into the prosodic consequences of rightward movement may help address this problem.

11. Thanks to Jeff Leopando, Lawrence Maligaya, and Joshua Monzon for participating in the study.

12. These examples contain the word *uláng* 'crayfish, lobster', which the speaker was unfamiliar with; he pronounced it *úlang*, with initial stress. In general, I will indicate stress where it was actually pronounced.

13. One might be excused for thinking that *nino* is a second-position clitic, attaching to the verb in (60a). Tagalog does have second-position clitics, but *nino* is not one of them. For one thing, it does not obey the general conditions on ordering of multiple second-position clitics. Bisyllabic second-position clitics of the NG class are required to precede those of the ANG class:

(i) a. Nakita nila   kayo (Tagalog)
        ACC-saw NG.they ANG.you.PL
        'They saw you'
    b. *Nakita kayo nila

Although it is of the NG class, *nino* follows all clitics, including those of the ANG class:

(ii) a. Nakita kayo   nino? (Tagalog)
         ACC-saw ANG.you.PL NG.who
         'Who saw you?'
     b. *Nakita nino kayo?

14. I have attempted to revise Simpson and Bhattacharya's transcription of Bangla to match Bayer's, hopefully not introducing too many errors in the process.

15. As Bayer and Simpson and Bhattacharya note, the same may be said to be true of Hindi, though the facts here are less clear, since Hindi tensed complement clauses must always be postverbal. The consequence is that although Hindi allows *wh* in situ in simple clauses (as in (i)), *wh* in situ in embedded clauses (as in (ii)) cannot take matrix scope:

(i) us-ne **kyaa** kiyaa? (Hindi)
     he-ERG what did
     'What did he do?'

(ii) a. *Raam-ne kahaa [ki **kOn** aayaa hE]?
         Raam-ERG said   that who come has
         'Who did Ram say has come?'
     b. **kOn** Raam-ne kahaa [ki __ aayaa hE]?

See Srivastav 1991 and Simpson 2000 for discussion.

16. Here, again, I have modified the transcription to match Bayer's.

17. Michaels and Nelson (2004), studying a different dialect of Bangla than Hayes and Lahiri do, report that their informant prefers to put focused phrases in immediately preverbal position. It is not clear how the strength of this preference compares with that in Basque, and I will have to leave this question for future investigation.

18. Many thanks to Lisa Selkirk for suggesting this move.

References

Abels, Klaus. 2003. *Successive cyclicity, anti-locality, and adposition stranding.* Doctoral dissertation, University of Connecticut.

Abney, Steven. 1987. *The English noun phrase in its sentential aspect.* Doctoral dissertation, MIT.

Ackema, Peter. 2001. Colliding complementizers in Dutch: Another syntactic OCP effect. *Linguistic Inquiry* 32:717–726.

Adger, David, and Daniel Harbour. 2007. Syntax and syncretisms of the Person Case Constraint. *Syntax* 10:2–37.

Aissen, Judith. 1979. *The syntax of causative constructions.* New York: Garland.

Aldridge, Edith. 2004. *Ergativity and word order in Austronesian languages.* Doctoral dissertation, Cornell University.

Alexiadou, Artemis. 2001. *Functional structure in nominals: Nominalization and ergativity.* Amsterdam: John Benjamins.

Alexiadou, Artemis, and Elena Anagnostopoulou. 2001. The subject in situ generalization and the role of case in driving computations. *Linguistic Inquiry* 32:193–231.

Alexiadou, Artemis, and Elena Anagnostopoulou. 2007. The subject in-situ generalization revisited. In Uli Sauerland and Hans-Martin Gärtner, eds., *Interfaces + recursion = language? Chomsky's minimalism and the view from syntax-semantics*, 31–60. Berlin: Walter de Gruyter.

Alexiadou, Artemis, Liliane Haegeman, and Melita Stavrou. 2007. *Noun phrase in the generative perspective.* Berlin: Mouton de Gruyter.

Anttila, Arto, and Vivienne Fong. 2000. The partitive constraint in Optimality Theory. *Journal of Semantics* 17:281–314.

Archangeli, Diana, and Douglas Pulleyblank. 2002. Kinande vowel harmony: Domains, grounded conditions, and one-sided alignment. *Phonology* 19:139–188.

Arregi, Karlos. 2002. *Focus on Basque movements.* Doctoral dissertation, MIT.

Ausín, Adolfo. 2000. A multiple spell-out account of *wanna*-contraction. Paper presented at WECOL.

Baker, Mark. 1988. *Incorporation.* Chicago: University of Chicago Press.

Baker, Mark. 2003. Agreement, dislocation, and partial configurationality. In Andrew Carnie, Heidi Harley, and MaryAnn Willie, eds., *Formal approaches to function in grammar: In honor of Eloise Jelinek*, 107–132. Philadelphia: John Benjamins.

Baker, Mark, and Chris Collins. 2006. Linkers and the internal structure of *v*P. *Natural Language and Linguistic Theory* 24:307–354.

Bammesberger, Alfred. 1983. *A handbook of Irish.* Heidelberg: Carl Winter Universitätsverlag.

Bayer, Josef. 1996. *Directionality and logical form: On the scope of focusing particles and wh-in-situ.* Dordrecht: Kluwer.

Bernstein, Judy. 1991. DPs in French and Walloon: Evidence for parametric variation in nominal head movement. *Probus* 3:101–126.

Bhatt, Rajesh, and Elena Anagnostopoulou. 1996. Object shift and specificity: Evidence from *ko*-phrases in Hindi. In Lise Dobrin, Kora Singer, and Lisa McNair, eds., *Proceedings of CLS 32*, 11–22. Chicago: Chicago Linguistic Society.

Bianchi, Valentina. 1999. *Consequences of antisymmetry: Headed relative clauses.* Berlin: Walter de Gruyter.

Biberauer, Theresa. 2008. Doubling and omission: Insights from Afrikaans negation. In Sjef Barbiers, Olaf Koeneman, Marika Lekakou, and Margreet van der Ham, eds., *Syntax and semantics, v. 36: Microvariation in syntactic doubling.* Bingley, UK: Emerald Group.

Bittner, Maria, and Ken Hale. 1996. Ergativity: Toward a theory of a heterogeneous class. *Linguistic Inquiry* 27:531–604.

Black, Cheryl. 1995. Boundary tones on word-internal domains in Kinande. *Phonology* 12:1–38.

Blaszczak, Joanna, and Hans-Martin Gärtner. 2005. Intonational phrasing, discontinuity, and the scope of negation. *Syntax* 8:1–22.

Bloomfield, Leonard. 1917. *Tagalog texts with grammatical analysis.* Urbana: University of Illinois.

Bobaljik, Jonathan, and Andrew Carnie. 1996. A minimalist approach to some problems of Irish word order. In Robert Borsley and Ian Roberts, eds., *The syntax of the Celtic languages*, 223–240. Cambridge: Cambridge University Press.

Bobaljik, Jonathan, and Idan Landau. 2009. Fact and fiction in Icelandic control. *Linguistic Inquiry* 40:113–132.

Bonet, Eulalia. 1991. *Morphology after Syntax: Pronominal Clitics in Romance Languages.* Doctoral dissertation, MIT.

Borsley, Robert, and Maria-Luisa Rivero. 1994. Clitic auxiliaries and incorporation in Polish. *Natural Language and Linguistic Theory* 12:373–423.

Bošković, Željko. 1997. On certain violations of the Superiority Condition, AgrO, and economy of derivation. *Journal of Linguistics* 33:227–254.

Bošković, Željko. 1998. LF movement and the Minimalist Program. In Pius Tamanji and Kiyomi Kusumoto, eds., *Proceedings of NELS 28*, 43–57. Amherst: GLSA, University of Massachusetts.

Bošković, Željko. 2001. *On the nature of the syntax-phonology interface: Cliticization and related phenomena.* Oxford: Elsevier Science.

Bresnan, Joan, and Jonni Kanerva. 1989. Locative inversion in Chicheŵa: A case study of factorization in grammar. *Linguistic Inquiry* 20:1–50.

Burzio, Luigi. 1986. *Italian syntax: A government-binding approach.* Dordrecht: Reidel.

Carnie, Andrew. 1995. *Non-verbal predication and head-movement.* Doctoral dissertation, MIT.

Cecchetto, Carlo, Carlo Geraci, and Sandro Zucchi. 2007. Another way to mark syntactic dependencies: The case for right peripheral specifiers in sign languages. Ms., Università degli Studi di Milano-Bicocca and Università degli Studi di Milano.

Chang, Lisa. 1997. *Wh-in-situ phenomena in French.* Master's thesis, University of British Columbia.

Cheng, Lisa, and Johan Rooryck. 2000. Licensing *wh*-in-situ. *Syntax* 3:1–19.

Chomsky, Noam. 1970. Remarks on nominalization. In Roderick Jacobs and Peter Rosenbaum, eds., *Readings in English transformational grammar*, 184–221. Waltham, MA: Ginn & Co.

Chomsky, Noam. 1977. On *wh* movement. In Peter Culicover, Thomas Wasow, and Adrian Akmajian, eds., *Formal Syntax*, 71–132. New York: Academic Press.

Chomsky, Noam. 1980. On binding. *Linguistic Inquiry* 11:1–46.

Chomsky, Noam. 1995. *The Minimalist Program*. Cambridge, MA: MIT Press.

Chomsky, Noam. 2000. Minimalist inquiries: The framework. In Roger Martin, David Michaels, and Juan Uriagereka, eds., *Step by step: Essays on minimalist syntax in honor of Howard Lasnik*, 89–155. Cambridge, MA: MIT Press.

Chomsky, Noam. 2001. Derivation by phase. In Michael Kenstowicz, ed., *Ken Hale: A life in language*, 1–52. Cambridge, MA: MIT Press.

Chomsky, Noam. 2005. On phases. Ms., MIT.

Chung, Sandra. 1990. VPs and verb movement in Chamorro. *Natural Language and Linguistic Theory* 8:559–619.

Cinque, Guglielmo. 1981. On the theory of relative clauses and markedness. *Linguistic Review* 1:247–294.

Collins, Chris. 2002. Eliminating labels. In Samuel David Epstein and T. Daniel Seely, eds., *Derivation and explanation in the Minimalist Program*, 42–64. Oxford: Blackwell.

Collins, Chris, and Phil Branigan. 1997. Quotative inversion. *Natural Language and Linguistic Theory* 15:1–41.

Comorovski, Ileana. 1996. *Interrogative phrases and the syntax-semantics interface*. Dordrecht: Kluwer.

Coon, Jessica. Forthcoming. VOS as predicate fronting in Chol. *Lingua*.

Davies, William, and Carol Rosen. 1988. Unions as multi-predicate clauses. *Language* 64:52–88.

Deguchi, Masanori, and Yoshihisa Kitagawa. 2002. Prosody and *wh*-questions. In Masako Hirotani, ed., *Proceedings of NELS 32*. Amherst: GLSA, University of Massachusetts.

Déprez, Viviane. 1988. Stylistic inversion and verb movement. In Joyce Powers and Kenneth de Jong, eds., *Proceedings of ESCOL '88*, pp. 71–82. Columbus: Ohio State University.

Dikken, Marcel den. 2006a. A reappraisal of *v*P being phasal—a reply to Legate. Ms., CUNY Graduate Center.

Dikken, Marcel den. 2006b. *Relators and linkers: The syntax of predication, predicate inversion, and copulas*. Cambridge, MA: MIT Press.

Dobashi, Yoshihito. 2004. Multiple spell-out, label-free syntax, and PF-interface. *Exploration in English Linguistics* 19.

Dobashi, Yoshihito. 2006. Cyclic spell-out, phonological phrasing, and focus. *Studies of Human Science* 2:65–75.

Doggett, Teal Bissell. 2004. *All things being unequal: Locality in movement*. Doctoral dissertation, MIT.

Downing, Laura. 2005. The prosody of focus-related enclitics in some Southern Bantu languages. Handout from talk given at SOAS.

Elordieta, Gorka. 1997. Accent, tone, and intonation in Lekeitio Basque. In Fernando Martínez-Gil and Alfonso Morales-Front, eds., *Issues in the phonology and morphology of the major Iberian languages*. Washington, DC: Georgetown University Press.

Embick, David. 1995. Mobile inflections in Polish. In Jill Beckman, ed., *Proceedings of NELS 25*, 127–142. Amherst: GLSA, University of Massachusetts.

Embick, David. 2000. Features, syntax, and categories in the Latin Perfect. *Linguistic Inquiry* 31:185–230.

Embick, David, and Rolf Noyer. 2006. Distributed Morphology and the syntax/morphology interface. In Gillian Ramchand and Charles Reiss, eds., *The Oxford Handbook of Linguistic Interfaces*, 289–324. Oxford: Oxford University Press.

Emonds, Joseph. 1976. *A transformational approach to English syntax*. New York: Academic Press.

Emonds, Joseph. 1985. *A unified theory of syntactic categories*. Dordrecht: Foris.

Endo, Yoshio. 1996. Right dislocation. In Masatoshi Koizumi, Masayuki Oishi, and Uli Sauerland, eds., *MITWPL 29: Formal approaches to Japanese linguistics 2*, 1–20. Cambridge, MA: MIT Working Papers in Linguistics.

Fabb, Nigel. 1984. *Syntactic affixation*. Doctoral dissertation, MIT.

Fanselow, Gisbert. 2001. Features, θ-roles, and free constituent order. *Linguistic Inquiry* 32:405–437.

Fox, Danny. 2000. *Economy and semantic interpretation*. Cambridge, MA: MIT Press.

Fox, Danny, and David Pesetsky. 2004. Cyclic linearization of syntactic structure. *Theoretical Linguistics* 31:1–46.

George, Leland. 1980. *Analogical generalizations of natural language syntax*. Doctoral dissertation, MIT.

Giorgi, Alessandra, and Giuseppe Longobardi. 1991. *The syntax of noun phrases: Configuration, parameters, and empty categories*. Cambridge: Cambridge University Press.

Grimshaw, Jane. 1997. The best clitic: Constraint conflict in morphosyntax. In Liliane Haegeman, ed., *Elements of grammar*, 169–196. Dordrecht: Kluwer.

Guasti, Maria Teresa. 1993. *Causative and perception verbs: A comparative study*. Torino, Italy: Rosenberg and Sellier.

Guasti, Maria Teresa. 1997. Romance causatives. In Liliane Haegeman, ed., *The new comparative syntax*, 124–144. London: Longman.

Guilfoyle, Eithne, Henrietta Hung, and Lisa Travis. 1992. Spec of IP and Spec of VP: Two subjects in Austronesian languages. *Natural Language and Linguistic Theory* 10:375–414.

Gussenhoven, Carlos. 2004. *The phonology of tone and intonation*. Cambridge: Cambridge University Press.

Hale, Ken. 2002. On the Dagur object relative: Some comparative notes. *Journal of East Asian Linguistics* 11:109–122.

Halle, Morris. 1990. An approach to morphology. In Juli Carter, Rose-Marie Dechaine, William Philip, and Tim Sherer, eds., *Proceedings of NELS 20*, 150–184. Amherst: GLSA, University of Massachusetts.

Halle, Morris, and Alec Marantz. 1993. Distributed Morphology and the pieces of inflection. In Kenneth Hale and S. Jay Keyser, eds., *The view from Building 20*, 111–176. Cambridge, MA: MIT Press.

Halpert, Claire. 2008. Subject position and agreement strategies in Kinande and Zulu. Handout from talk given at MIT.

Hamlaoui, Fatima. 2008. On the role of discourse and phonology in French *wh*-questions. Ms., University of Ottawa.

Harada, Shin-ichi. 1973. Counter Equi-NP deletion. *Annual Bulletin* 7:113–148. Tokyo: Research Institute of Logopedics and Phoniatrics, Tokyo University.

Harbour, Daniel. 2007. A feature calculus for Silverstein hierarchies. Handout from talk given at the Workshop on Morphology and Argument Encoding, Harvard University, September 8.

Harley, Heidi. 1995. *Subjects, events, and licensing*. Doctoral dissertation, MIT.

Hayes, Bruce, and Aditi Lahiri. 1991. Bengali intonational phonology. *Natural Language and Linguistic Theory* 9:47–96.

Hinnebusch, Thomas, and Sarah Mirza. 1979. *Kiswahili: Msingi wa kuseme kusoma na kuandika / Swahili: A foundation for speaking, reading, and writing*. Lanham, MD: University Press of America.

Hiraiwa, Ken. Forthcoming. Spelling out the double-*o* constraint. *Natural Language and Linguistic Theory*.

Hiraiwa, Ken, and Shinichiro Ishihara. 2002. Missing links: Cleft, sluicing, and *"no da"* construction in Japanese. In Tania Ionin, Heejeong Ko, and Andrew Nevins, eds., *MITWPL 43: Proceedings of HUMIT 2001*, 35–54. Cambridge, MA: MIT Working Papers in Linguistics.

Hirotani, Masako. 2005. *Prosody and LF interpretation: Processing Japanese wh-questions*. Doctoral dissertation, University of Massachusetts, Amherst.

Hoekstra, Teun. 1984. *Transitivity*. Dordrecht: Foris.

Iatridou, Sabine, and Hedde Zeijlstra. 2009. The need for differentiation. Class handout, MIT.

Igarashi, Yosuke, and Yoshihisa Kitagawa. 2007. Focus and Monotony. Handout for talk given at the 3rd Workshop on Prosody, Syntax, and Information Structure, Indiana University, September 14–15, 2007.

Ishihara, Shinichiro. 2001. Stress, focus, and scrambling in Japanese. In Elena Guerzoni and Ora Matushansky, eds., *MITWPL 39: A few from Building E39*, 142–175. Cambridge, MA: MIT Working Papers in Linguistics.

Ishihara, Shinichiro. 2003. *Intonation and interface conditions*. Doctoral dissertation, MIT.

Ishihara, Shinichiro. 2007. Major Phrase, focus intonation, multiple spell-out (MaP, FI, MSO). *Linguistic Review* 24:137–167.

Ito, Junko, and Armin Mester. 2007. Prosodic categories and recursion. Talk given at WPSI 3, Indiana University.

Jones, Patrick, and Jessica Coon. 2008. Tone in Kinande. Ms., MIT.

Kahnemuyipour, Arsalan. 2005. Towards a phase-based theory of sentential stress. In Martha McGinnis and Norvin Richards, eds., *MITWPL 49: Perspectives on phases*, 125–146. Cambridge, MA: MIT Working Papers in Linguistics.

Kanerva, Jonni. 1989. *Focus and phrasing in Chichewa phonology*. Doctoral dissertation, Stanford University.

Kanerva, Jonni. 1990. Focusing on phonological phrases in Chichewa. In Sharon Inkelas and Draga Zec, eds., *The phonology-syntax connection*. Chicago: University of Chicago Press.

Kaufman, Daniel. 2005. Aspects of pragmatic focus in Tagalog. In I Wayan Arka and Malcom Ross, eds., *The many faces of Austronesian voice systems: Some new empirical studies*, 175–196. Canberra: Pacific Linguistics.

Kayne, Richard. 1972. Subject inversion in French interrogatives. In Jean Casagrande and Bohdan Saciuk, eds., *Generative studies in Romance languages*, 70–126. Rowley, MA: Newbury House.

Kayne, Richard. 1975. *French Syntax: The Transformational Cycle*. Cambridge, MA: MIT Press.

Kayne, Richard. 1977. French relative *que*. In Marta Luján and Fritz Hensey, eds., *Current Studies in Romance Linguistics*. Washington, DC: Georgetown University Press.

Kayne, Richard. 1984. *Connectedness and binary branching*. Dordrecht: Foris.

Kayne, Richard. 1994. *Antisymmetry*. Cambridge, MA: MIT Press.

Kayne, Richard. 2004. Prepositions as probes. In Adriana Belletti, ed., *Structures and beyond: The cartography of syntactic structures*, vol. 3. Oxford: Oxford University Press.

Kayne, Richard, and Jean-Yves Pollock. 1978. Stylistic inversion, successive cyclicity, and move NP in French. *Linguistic Inquiry* 9:595–621.

Koizumi, Masatoshi. 1993. Object agreement phrases and the split VP hypothesis. In Jonathan Bobaljik and Colin Phillips, eds., *MITWPL 18: Case and agreement I*, 99–148. Cambridge, MA: MIT Working Papers in Linguistics.

Koizumi, Masatoshi. 2000. String vacuous overt verb raising. *Journal of East Asian Linguistics* 9:227–285.

Kornfilt, Jaklin. 1986. The Stuttering Prohibition and morpheme deletion in Turkish. In Ayhan Aksu Koç and Eser Erguvanlı Taylan, eds., *Proceedings of the Turkish Linguistics Conference*, 59–83. Istanbul: Boğaziçi University Publications.

Kratzer, Angelika, and Elisabeth Selkirk. 2007. Phase theory and prosodic spellout: The case of verbs. *Linguistic Review* 24:93–135.

Krause, Cornelia. 2001. *On reduced relatives with genitive subjects.* Doctoral dissertation, MIT.

Kubozono, Haruo. 2007. Focus and intonation in Japanese: Does focus trigger pitch reset? In S. Ishihara, S. Jannedy, and A. Schwartz., eds., *Interdisciplinary studies on information structure* 9:1–27. Potsdam: Universitätsverlag Potsdam.

Kuroda, Shige-Yuki. 1988. Whether we agree or not: A comparative syntax of English and Japanese. *Linguisticae Investigationes* 12:1–47.

Kuwabara, Kazuki. 1996. Multiple *wh*-phrases in elliptical clauses and some aspects of clefts with multiple foci. In Masatoshi Koizumi, Masayuki Oishi, and Uli Sauerland, eds., *MITWPL 29: Formal approaches to Japanese linguistics 2*, 97–116. Cambridge, MA: MIT Working Papers in Linguistics.

Landau, Idan. 2004. Severing the distribution of PRO from Case. *Syntax* 9:153–170.

Lee, Felicia. 2000. VP remnant movement and VSO in Quiavini Zapotec. In Andrew Carnie and Eithne Guilfoyle, eds., *The syntax of verb-initial languages*, 143–162. Oxford: Oxford University Press.

Legate, Julie. 2003. Some interface properties of the phase. *Linguistic Inquiry* 34:506–515.

Longobardi, Giuseppe. 1980. Remarks on infinitives: A case for a filter. *Journal of Italian Linguistics* 1(2):101–155.

Longobardi, Giuseppe. 1994. Reference and proper names: A theory of N-movement in syntax and Logical Form. *Linguistic Inquiry* 25:609–665.

Longobardi, Giuseppe. 2001. The structure of DPs: Some principles, parameters, and problems. In Mark Baltin and Chris Collins, eds., *The handbook of contemporary syntactic theory*, 562–603. Oxford: Blackwell.

Marantz, Alec. 1991. Case and licensing. In Germán Westphal, Benjamin Ao, and Hee-Rahk Chae, eds., *Proceedings of ESCOL '91*, 234–253. Columbus: Ohio State University.

Marantz, Alec. 1993. Implications and asymmetries in double object constructions. In Sam Mchombo, ed., *Theoretical aspects of Bantu grammar I*. Stanford, CA: CSLI.

Marantz, Alec. 1997. No escape from syntax: Don't try morphological analysis in the privacy of your own lexicon. In Alexis Dimitriadis et al., eds., *Proceedings of the 21st Annual Penn Linguistics Colloquium: Penn Working Papers in Linguistics 4:2*, 201–225.

Massam, Diane. 1985. *Case theory and the projection principle.* Doctoral dissertation, MIT.

Massam, Diane. 2000. VSO and VOS: Aspects of Niuean word order. In Andrew Carnie and Eithne Guilfoyle, eds., *The syntax of verb-initial languages*, 97–116. Oxford: Oxford University Press.

Massam, Diane. 2001. Pseudo noun incorporation in Niuean. *Natural Language and Linguistic Theory* 19:153–197.

McCloskey, James. 1996. Subjects and subject positions in Irish. In Robert Borsley and Ian Roberts, ed., *The syntax of the Celtic languages*, 241–283. Cambridge: Cambridge University Press.

McGinnis, Martha. 1998. *Locality in A-movement.* Doctoral dissertation, MIT.

McGinnis, Martha. 2001. Variation in the phase structure of applicatives. In Johan Rooryck and Pierre Pica, eds., *Linguistic variations yearbook*. Amsterdam: John Benjamins.

Menn, Lise, and Brian MacWhinney. 1984. The Repeated Morph Constraint: Toward an explanation. *Language* 60:519–541.

Merchant, Jason. 2001. *The syntax of silence: Sluicing, islands, and the theory of ellipsis.* Oxford: Oxford University Press.

Michaels, Jennifer, and Catherine Nelson. 2004. A preliminary investigation of intonation in East Bengali. Ms., UCLA.

Miyagawa, Shigeru. Forthcoming. *Why move? Why agree?* Cambridge, MA: MIT Press.

Mohanan, Tara. 1994a. *Argument structure in Hindi.* Stanford, CA: CSLI.

Mohanan, Tara. 1994b. Case OCP: A constraint on word order in Hindi. In Miriam Butt, Tracy Holloway King, and Gillian Ramchand, eds., *Theoretical perspectives on word order in South Asian languages*, 185–216. Stanford, CA: CSLI.

Moltmann, Friederike. 1995. Exception sentences and polyadic quantification. *Linguistics and Philosophy* 18:223–280.

Moro, Andrea. 2000. *Dynamic antisymmetry.* Cambridge, MA: MIT Press.

Munn, Alan. 1995. The possessor that stayed close to home. In Vida Samiian and Jeanette Schaeffer, eds., *Proceedings of WECOL 24*, 181–195.

Murayama, Kazuto. 1998. An argument for Japanese Right Dislocation as feature-driven movement. Ms., Kanda University of International Studies.

Mutaka, Ngessimo. 1994. *The lexical tonology of Kinande.* Munich: LINCOM Europa.

Mutaka, Ngessimo, and Kambale Kavutirwaki. To appear. Kinande/Konzo-English Dictionary with an English-Kinande/Konzo Index. Brussels: Tervuren.

Ndayiragije, Juvenal. 1999. Checking economy. *Linguistic Inquiry* 30:399–444.

Neeleman, Ad, and Hans van de Koot. 2005. Syntactic OCP effects. In Martin Everaert and Henk van Riemsdijk, eds., *The Blackwell syntactic companion.* London: Blackwell.

Neidle, Carol, Judy Kegl, Dawn MacLaughlin, Benjamin Bahan, and Robert G. Lee. 2000. *The syntax of American Sign Language: Functional categories and hierarchical structure.* Cambridge, MA: MIT Press.

Nespor, Marina, and Irene Vogel. 1986. *Prosodic phonology.* Dordrecht: Foris.

Nevins, Andrew. 2004. Derivations without the Activity Condition. In Martha McGinnis and Norvin Richards, eds., *MITWPL 49: Perspectives on phases*, 287–310. Cambridge, MA: MIT Working Papers in Linguistics.

Nissenbaum, Jon. 2000. *Investigations of covert phrase movement.* Doctoral dissertation, MIT.

Odden, David. 1996. *The phonology and morphology of Kimatuumbi.* Oxford: Oxford University Press.

Pak, Marjorie. 2008. *The postsyntactic derivation and its phonological reflexes.* Doctoral dissertation, University of Pennsylvania.

Perlmutter, David. 1971. *Deep and Surface Structure Constraints in Syntax.* New York: Holt, Rinehart & Winston.

Pesetsky, David. 1987. *Wh*-in-situ: Movement and unselective binding. In Eric Reuland and Alice ter Meulen, eds., *The representation of (in)definiteness*, 98–129. Cambridge, MA: MIT Press.

Pesetsky, David. 1995. *Zero syntax.* Cambridge, MA: MIT Press.

Pesetsky, David. 1998. Some optimality principles of sentence pronunciation. In Pilar Barbosa, Danny Fox, Paul Hagstrom, Martha McGinnis, and David Pesetsky, eds., *Is the best good enough?: Optimality and competition in syntax.* Cambridge, MA: MIT Press and MITWPL.

Pesetsky, David. 2000. *Phrasal movement and its kin.* Cambridge, MA: MIT Press.

Pesetsky, David, and Esther Torrego. 2006. Probes, goals, and syntactic categories. In Yukio Otsu, ed., *Proceedings of the 7th annual Tokyo Conference on Psycholinguistics*. Tokyo: Hituzi Syobo Publishing Company.

Pierrehumbert, Janet, and Mary Beckman. 1988. *Japanese tone structure*. Cambridge, MA: MIT Press.

Pires, Acrisio, and Heather Taylor. 2007. The syntax of *wh*-in-situ and common ground. Submitted to *Proceedings of the 37th LSRL*. Amsterdam: John Benjamins.

Poser, William. 1984. *The phonetics and phonology of tone and intonation in Japanese*. Doctoral dissertation, MIT.

Poser, William. 2002. The double-o constraints in Japanese. Ms., University of Pennsylvania.

Pylkkänen, Liina. 2002. *Introducing arguments*. Doctoral dissertation, MIT.

Rackowski, Andrea. 2002. *The structure of Tagalog: Specificity, voice, and the distribution of arguments*. Doctoral dissertation, MIT.

Rackowski, Andrea, and Lisa Travis. 2000. V-initial languages: X or XP movement and adverbial placement. In Andrew Carnie and Eithne Guilfoyle, eds., *The syntax of verb-initial languages*, 117–141. Oxford: Oxford University Press.

Raimy, Eric. 2000. *The phonology and morphology of reduplication*. Berlin: Mouton de Gruyter.

Reglero, Lara. 2005. *Wh*-in-situ constructions: Syntax and/or phonology? In John Alderete, Chung-Hye Han and Alexei Kochetov́, eds., *Proceedings of WCCFL 24*, 334–342. Somerville, MA: Cascadilla Press.

Richards, Marc. 2004. *Object shift and scrambling in North and West Germanic: A case study in symmetrical syntax*. Doctoral dissertation, University of Cambridge.

Richards, Marc. 2007. On feature inheritance: An argument from the Phase Impenetrability Condition. *Linguistic Inquiry* 38:563–572.

Richards, Norvin. 1993. *Tagalog and the typology of scrambling*. Honors thesis, Cornell University.

Richards, Norvin. 1999. Featural cyclicity and the ordering of multiple specifiers. In Samuel D. Epstein and Howard Lasnik, eds., *Working Minimalism*, 127–158. Cambridge, MA: MIT Press.

Richards, Norvin. 2000. Another look at Tagalog subjects. In Ileana Paul, Viviane Phillips, and Lisa Travis, eds., *Formal issues in Austronesian linguistics*. Dordrecht: Kluwer.

Richards, Norvin. 2001. *Movement in language: Interactions and architectures*. Oxford: Oxford University Press.

Ritter, Elizabeth. 1991. Two functional categories in noun phrases: Evidence from modern Hebrew. In Susan Rothstein, ed., *Syntax and semantics 25: Perspectives on phrase structure*, 37–62. New York: Academic Press.

Rizzi, Luigi. 1997. The fine structure of the left periphery. In Liliane Haegeman, ed., *Elements of grammar: Handbook in generative syntax*, 281–337. Dordrecht: Kluwer.

Rodríguez-Mondoñedo, Miguel. 2007. *The syntax of objects: Agree and differential object marking*. Doctoral dissertation, University of Connecticut.

Ross, John. 1972. Doubl-ing. *Linguistic Inquiry* 3:61–86.

Rouveret, Alain, and Jean-Roger Vergnaud. 1980. Specifying reference to the subject: French causatives and conditions on representations. *Linguistic Inquiry* 11:97–202.

Rudin, Catherine. 1985. *Aspects of Bulgarian syntax: Complementizers and WH constructions*. Columbus, OH: Slavica.

Rudin, Catherine. 1988. On multiple questions and multiple *wh*-fronting. *Natural Language and Linguistic Theory* 6:445–501.

Sabbagh, Joseph. 2008. Copular clauses and predicate initial word order in Tagalog. Ms., Reed College.

Saito, Mamoru. 2002. On the role of selection in the application of Merge. In Makoto Kadowaki and Shigeto Kawahara, eds., *Proceedings of NELS 33*, 323–346. Amherst: GLSA, University of Massachusetts.

Salanova, Andres. 2004. Obligatory inversion as avoidance of unlinearizable representations. Ms., MIT.

Sandalo, Filomena, and Hubert Truckenbrodt. 2002. Some notes on phonological phrasing in Brazilian Portuguese. In Aniko Csirmaz, Zhiqiang Li, Andrew Nevins, Olga Vaysman, and Michael Wagner, eds., *MITWPL 42: Phonological answers (and their corresponding questions)*, 285–310. Cambridge, MA: MIT Working Papers in Linguistics.

Santorini, Beatrice, and Caroline Heycock. 1988. *Remarks on causatives and passive*. Technical Report No. MS-CIS-88-33. Philadelphia: Department of Computer and Information Science, University of Pennsylvania.

Sauerland, Uli. 1995. Sluicing and islands. Ms., MIT.

Schachter, Paul. 1976. The subject in Philippine languages: Actor, topic, actor-topic, or none of the above. In Charles Li, ed., *Subject and Topic*, 491–518. New York: Academic Press.

Schachter, Paul. 1996. The subject in Tagalog: Still none of the above. *UCLA Occasional Papers in Linguistics* 15.

Schachter, Paul, and Fe Otanes. 1972. *Tagalog reference grammar*. Berkeley: University of California Press.

Schneider-Zioga, Patricia. 2000. Anti-agreement and the fine structure of the left periphery. In Ruixi Ai, Francesca del Gobbo, Makie Irie, and Hajime Ono, eds., *UCI Working Papers in Linguistics* 6, 94–114. Irvine: Department of Linguistics, University of California.

Schneider-Zioga, Patricia. 2007. Anti-agreement, antilocality, and minimality: The syntax of dislocated subjects. *Natural Language and Linguistic Theory* 25:403–446.

Schütze, Carson. 1997. *INFL in child and adult language: Agreement, case, and licensing*. Doctoral dissertation, MIT.

Schwartz, Florian. 2007. *Ex-situ* focus in Kikuyu. In Enoch Oladé Aboh, Katharina Hartmann, and Malte Zimmermann, eds., *Focus strategies in African languages: The interaction of focus and grammar in Niger-Congo and Afro-Asiatic*, 139–160. Berlin: Walter de Gruyter.

Seidl, Amanda. 2001. *Minimal indirect reference: A theory of the syntax-phonology interface*. New York: Routledge.

Selkirk, Elisabeth. 1984. *Phonology and syntax: The relation between sound and structure*. Cambridge, MA: MIT Press.

Selkirk, Elisabeth. 1986. On derived domains in sentence phonology. *Phonology Yearbook* 3:371–405.

Selkirk, Elisabeth. 1995. Sentence prosody: Intonation, stress, and phrasing. In John Goldsmith, ed., *The handbook of phonological theory*, 550–569. Cambridge: Blackwell.

Selkirk, Elisabeth. 2006. Bengali intonation revisited: An Optimality Theoretic analysis in which FOCUS stress prominence drives FOCUS phrasing. In Chungmin Lee, Matthew Gordon, and Daniel Büring, eds., *Topic and focus, intonation and meaning*, 215–244. Dordrecht: Springer.

Selkirk, Elisabeth. 2009. Spelling out syntactic constituents as prosodic domains: Match constraints and the syntax-prosodic structure interface. Handout from talk given at Harvard University.

Selkirk, Elisabeth, and Koichi Tateishi. 1988. Constraints on minor phrase formation in Japanese. In Lynn MacLeod, Gary Larson, and Diane Brentari, eds., *Proceedings of CLS 24*, 316–336. Chicago, IL: Chicago Linguistic Society.

Selkirk, Elisabeth, and Koichi Tateishi. 1991. Syntax and downstep in Japanese. In Carol Georgopoulos and Roberta Ishihara, eds., *Interdisciplinary approaches to language: Essays in honor of S.-Y. Kuroda*. Dordrecht: Kluwer.

Sigurðsson, Halldór. 1991. Icelandic case-marked PRO and the licensing of lexical arguments. *Natural Language and Linguistic Theory* 9:327–363.

Silverstein, Michael. 1976. Hierarchy of features and ergativity. In Robert Dixon, ed., *Grammatical categories in Australian languages*. Canberra: Australian Institute of Aboriginal Studies.

Simpson, Andrew. 2000. *Wh-movement and the theory of feature checking*. Amsterdam: John Benjamins.

Simpson, Andrew, and Tanmoy Bhattacharya. 2003. Obligatory overt *wh*-movement in a *wh*-in-situ language. *Linguistic Inquiry* 34:127–142.

Skopeteas, Stavros, and Elisabeth Verhoeven. 2009. Distinctness effects on VOS order: Evidence from Yucatec Maya. In Heriberto Avelino, Jessica Coon, and Elisabeth Norcliffe, eds., MITWPL 59: pp. 157–173. *New perspectives in Mayan linguistics*. Cambridge, MA: MIT Working Papers in Linguistics.

Smith, Jennifer. 2005. On the *wh*-question intonational domain in Fukuoka Japanese: Some implications for the syntax-prosody interface. In Shigeto Kawahara, ed., *UMOP 30: Papers on prosody*. Amherst: GLSA, University of Massachusetts.

Soh, Hooi Ling. 1998. *Object scrambling in Chinese*. Doctoral dissertation, MIT.

Srivastav, Veneeta. 1991. Subjacency effects at LF: The case of Hindi wh. *Linguistic Inquiry* 22:762–769.

Stowell, Tim. 1981. *Origins of phrase structure*. Doctoral dissertation, MIT.

Strozer, Judith. 1976. *Clitics in Spanish*. Doctoral dissertation, UCLA.

Sugahara, Mariko. 2003. *Downtrends and post-FOCUS intonation in Tokyo Japanese*. Doctoral dissertation, University of Massachusetts, Amherst.

Szabolcsi, Anna. 1994. The noun phrase. In Ferenc Kiefer and Katalin É. Kiss, eds., *Syntax and semantics 27: Syntactic structure of Hungarian*. New York: Academic Press, pp. 179–275.

Szczegielniak, Adam. 1997. Deficient heads and long head movement in Slovak. In Martin Lindseth and Steven Franks, eds., *Proceedings of FASL 5, The Indiana Meeting*. Ann Arbor: Michigan Slavic Publications, pp. 312–333.

Szczegielniak, Adam. 1999. "That-t effects" crosslinguistically and successive cyclic movement. In Karlos Arregi, Benjamin Bruening, Cornelia Krause, and Vivian Lin, eds., *MITWPL 33: Papers on morphology and syntax, cycle one*, 369–393. Cambridge, MA: MIT Working Papers in Linguistics.

Szendrői, Kriszta. 2001. *Focus and the syntax-phonology interface*. Doctoral dissertation, University College London.

Takahashi, Daiko. 1993. Movement of *wh*-phrases in Japanese. *Natural Language and Linguistic Theory* 11:655–678.

Takahashi, Daiko. 1994. Sluicing in Japanese. *Journal of East Asian Linguistics* 3:265–300.

Torrego, Esther. 1998. *The dependencies of objects*. Cambridge, MA: MIT Press.

Truckenbrodt, Hubert. 1995. *Phonological phrases: Their relation to syntax, focus, and prominence*. Doctoral dissertation, MIT.

Truckenbrodt, Hubert. 1999. On the relation between syntactic phrases and phonological phrases. *Linguistic Inquiry* 30:219–255.

Uechi, Akihiko. 1998. *An interface approach to topic/focus structure*. Doctoral dissertation, University of British Columbia.

Uribe-Etxebarria, Myriam. 2002. In situ questions and masked movement. In P. Pica and J. Rooryck, eds., *Linguistic Variation Yearbook*. Amsterdam: John Benjamins.

Valois, Daniel. 1991. The internal structure of DP and adjective placement in French and English. In Tim Sherer, ed., *Proceedings of NELS 21*, 367–382. Amherst: GLSA, University of Massachusetts.

Valois, Daniel, and Fernande Dupuis. 1992. On the status of (verbal) traces in French: The case of stylistic inversion. In Paul Hirschbühler and Konrad Koerner, eds., *Romance languages and modern linguistic theory*, 325–338. Amsterdam: John Benjamins.

Vicente, Luis. 2005. Word order variation in Basque as non-feature-driven movement. Ms., Leiden University.

Wagner, Michael. 2005. *Prosody and recursion*. Doctoral dissertation, MIT.

Walter, Mary Anne. 2005. The Distinctness Condition in Semitic syntax. Ms., MIT.

Wu, Hsiao-Hung Iris. 2008. *Generalized inversion and the theory of Agree*. Doctoral dissertation, MIT.

Wurmbrand, Susi. 1998. Downsizing infinitives. In Uli Sauerland and Orin Percus, eds., *MITWPL 25: The interpretive tract*, 141–175. Cambridge, MA: MIT Working Papers in Linguistics.

Wurmbrand, Susi. 2003. *Infinitives: Restructuring and clause structure*. Berlin: Walter de Gruyter.

Wurmbrand, Susi. 2006. Licensing Case. *Journal of Germanic Linguistics* 18:175–236.

Yang, Dong-Whee. 2005. Focus movements, Distinctness condition, and intervention effects. Ms, MIT.

Yip, Moira. 1998. Identity avoidance in phonology and morphology. In Steven Lapointe, Diane Brentari, and Patrick Farrell, eds., *Morphology and its relation to phonology and syntax*. Stanford, CA: CSLI.

Zubizarreta, Maria Luisa. 1998. *Prosody, focus, and word order*. Cambridge, MA: MIT Press.

Index

insertion
    early   210
    interface(s) 219 (Embick & N. 2006)

[handwritten margin notes: "phase 200–" next to Phonologically null; "retraction – SEE stress / tone"; "reversal – SEE subject"; "spell-out 201–3 206" near Pulleyblank/Pylkkänen; "sign lg. SEE Italian" next to Shortest Attract; "SEE inversion" in right margin]

vocabulary $6^2, 7^4$ $\underline{70}$

Linguistic Inquiry Monographs

Samuel Jay Keyser, general editor